Can Science Save Us?

Other Books by
GEORGE A. LUNDBERG

Foundations of Sociology
Social Research, 2nd ed.
Leisure: A Suburban Study
1933)
Trends in American Sociology
Sociology, 2nd ed.

Can Science Save Us?

by
GEORGE A. LUNDBERG
UNIVERSITY OF WASHINGTON

Second Edition

GREENWOOD PRESS, PUBLISHERS
WESTPORT, CONNECTICUT

Library of Congress Cataloging in Publication Data

Lundberg, George Andrew, 1895-1966.
Can science save us?

Reprint of the 2d. ed. published by Longmans, Green, New York.
Includes index.
1. Science--Social aspects. I. Title.
[Q175.5.L86 1980] 301.24'3 79-16792
ISBN 0-313-21299-6

Originally published by Longmans, Green and Co., Inc., New York
First edition February 1947
Second edition April 1961

Reprinted with the permission of Longman Inc., New York

Reprinted in 1979 by Greenwood Press, Inc.
51 Riverside Avenue, Westport, Conn. 06880

Printed in the United States of America

10 9 8 7 6 5 4 3 2 1

Contents

CONTENTS

Can Science Save Us?

CHAPTER I

Prescientific Thoughtways in a Technological Age

I

Science Produces Social Problems

THE 1960's began with history's greatest fanfare for science. Man-made satellites whirled their way in the earth's orbit sending back information of regions and phenomena hitherto unknown. Qualified experts agree that man himself will presently navigate these realms, and far beyond. In the meantime, comparable developments in missiles and other deadly weapons raise the specter of possible annihilation of the generation who engineered these marvels. Will this same generation or its successors (if any) be able to develop sciences of human social relations adequate to manage and to utilize constructively the forces and instruments now for the first time in the hands of men? Faced with this question, even physical scientists, appalled at the possible social consequences of their handiwork, sometimes join in the periodic hue and cry for a "moratorium" on physical science. The course of scientific advance meanwhile plunges forward, for better or for worse.

Actually, very few people seriously question the importance and the promise of the advancement of science. Nearly everybody agrees that, whatever other troubles the advancement of science may have brought in its wake, it has released us from some age-long fears and insecurities. The natural sciences have undoubtedly given us a large measure of control over many of our traditional enemies, although we may not always exercise this power.

Even when we cannot do anything directly about averting natural events, science is still invaluable in two principal ways: First, science forewarns us of certain events and thus enables us to avoid their more serious consequences. If rain is predicted, we carry an unbrella. Ships are warned that a hurricane is likely to follow a certain path. We may protect ourselves from certain diseases by proper inoculations; and so forth. Second, the mere possession of scientific knowledge and scientific habits of thought regarding the natural universe relieves us of a world of fears, rages, and other unpleasant dissipations of energy. Scientific knowledge operates as a sort of mental hygiene in the fields where it is applied. If the morning paper reports an earthquake, an eclipse, a storm, or a flood, these events are immediately referred to their proper place in the framework of science, in which their explanation, i.e., their relationship to other events, has already been worked out. Hence, each new event of this character calls for very little, if any, "mental" or "emotional" strain upon the organism so far as our intellectual adjustment to it as an event is concerned.

Political and social upheavals, on the other hand, such as wars, revolutions, and crime are to most people a matter of shock and much personal recrimination and other emotionalism. The nervous condition of most commentators on the current social situation is not conducive either to the clearest kind of analysis or to good digestion. Floods, hurricanes, and crop failures also worry us, but we take a very different type of attitude toward them than we take toward social misfortunes.

It is not surprising, therefore, that man should have become increasingly interested in the question of whether this same tool called science might not be useful to him also in his social predicaments. On this question, however, the sharpest difference of opinion exists. Many thoughtful people hold that the advancement of science, far from alleviating social problems, has greatly aggravated them. "The world," says Robert Hutchins, "has reached at one and the same moment the zenith of its information, technology, and power over nature and the nadir of its moral and political life."

Let us neglect for the time being whether this is an overstatement of fact. Let us also pass over the logical implication that, if two events occur together or in sequence, one is necessarily the cause of the other. It remains true that, whether or not our social relations are at an all-time low at present, scientific advancement in physical and biological fields has not resolved the really outstanding problems of our age, namely, the problems of human relations.

It simply cannot be shown that improvements in industrial products and in the purchasing power of workers have resulted in more peaceful and friendly employer-employee relationships. The mere rise in standards and planes of living, better health, and abundant crops have not made communities or nations peaceful and friendly. It does not follow that these developments are not in themselves admirable. But we must admit that they do not *of themselves* provide the road to peace and amicable human relations.

Where, then, shall we turn for help regarding social problems?

II

Why Not Try Social Science?

In his long career on earth, man has tried with varying degrees of success numerous methods of coming to terms with his world. It is unnecessary to review the story. Its main outlines are familiar. Suffice it to say that the scientific method as developed in the physical sciences during the last four hundred years has proved incomparably powerful in solving certain age-long problems of cold, darkness, famine, epidemics, distance, communication, transportation, and a thousand other needs of man throughout the centuries. Ironically, he finds himself today, perhaps largely as a result of his technological triumphs, engulfed in difficulties with his fellow men which threaten to despoil him of his enjoyment of the choicest fruits that science has placed within his reach.

Why does he not turn in this predicament to the methods which have proved themselves so potent in other fields?

The principal reason is simple tradition. Human relations are not yet generally believed to be proper subjects for serious scientific study. The conspicuous triumphs of science thus far have been in such fields as health, fertilizer, and improved mechanical gadgets. These are, therefore, deemed the only proper provinces of scientific research.

To be sure, there is much talk about "social science," scientific attitudes in human affairs, and various "scientific" reform movements. But a little examination will usually reveal that nobody, including many so-called social scientists, really takes science seriously in such connections. The word *science* has mainly honorific significance as applied to human relations. Science means to most people a remarkably reliable magic in certain physical fields. As such it is a prestige word of great potency. It is therefore attached to all manner of social programs, in order that they may shine by the reflected glory of the neon light and the radio tube. Indeed, a great many accredited social scientists, in the sense of practicing economists, sociologists, psychologists, anthropologists, and above all "political" scientists, are not themselves convinced that science can and should mean, in their fields, the same impersonal, rigorous, and nonethical discipline that the word implies in the physical world. Science is a relatively new thoughtway, and the idea of looking to it for solutions of problems of human relations has not yet penetrated very far. The dominant thoughtway of our time as regards social problems is a legalistic-moralistic, "literary" orientation which has been largely abandoned in coming to terms with our physical environment. This ancient type of thinking has deep intrenchments in the emotional and esthetic feelings of people but it does not yield results of the kind sought by science. We shall return to this subject in a later chapter. (See pp. 66–67.)

A second reason why interest in the advancement of the social sciences is not greater is that most people feel they already know the answers to problems of human relations. What are some of these solutions? Why—"an honest day's work for an honest day's pay"; "the Ten Commandments"; "the Golden Rule"; socialism; free enterprise; communism; cooperation; another World Conference; World Union; and so forth. This idea that the expression of an ideal constitutes also a means for its attainment is simply a survival of magical thinking. Science is not a substitute for ideals. It is the most effective instrument of their attainment. The fact is, we do not know how feasibile or effective any of these ideal programs might be under various conditions. Neither do we know how to estimate or demonstrate their feasibility and value. To discover and demonstrate these things is one purpose of research in the social sciences.

Illustrations of how ignorance and vindictiveness, born of prescientific modes of thought, have operated disastrously in our management of international affairs may be drawn from the peace settlements following the two world wars. Allowing for political and other pressures under which the peacemakers were laboring, the fact remain that all of the *principals* of the peace conferences were ignorant about the kind of knowledge that mattered. Speaking about the settlement after World War I, H. G. Wells gives the following appraisal: "These people don't know anything about the business they have in hand. Nobody knows very much, but the important thing to realize is that they do not even know what is to be known. They arrange so and so, and so and so must ensue, and they cannot or will not see that so and so must ensue. They are so unaccustomed to competent thought, so ignorant that there is knowledge, and of what knowledge is, that they do not understand that it matters."

Mr. Wells goes on to remark on this state of affairs in the following words: "The same terrifying sense of insufficient mental equipment was dawning upon some of us who watched the birth of the League of Nations. Reluctantly and with something like horror, we realized that these people who were, they imagined, turning over a new page and beginning a fresh chapter in human history, knew collectively hardly anything about the formative forces in history. Collectively I say. Altogether they had a very considerable amount of knowledge, uncoordinated bits of quite good knowledge, some about this period and some about that, but they had no common understanding whatever of the processes in which they were obliged to mingle and interfere.

"If I might attempt a sweeping generalization about the general course of human history in the eighteen years that have followed the War, I would describe it as a series of flounderings, violent, ill-directed mass movements, slack drifting here, and convulsive action there. . . . And it is so largely tragic because the creature really is intelligent, can feel finely and acutely, express itself poignantly in art, music, and literature, and—this is what I am driving at—impotently knows better."[1]

[1] H. G. Wells, "The Idea of a World Encyclopedia," *Harper's Magazine,* April, 1937.

A decade later, one of our most thoughtful diplomats, a career man of twenty-five years' standing in our State Department comments as follows:

"I should like first to say a word about the total result of these two world wars in Europe. These wars were fought at the price of some tens of millions of lives, of untold physical destruction of the balance of forces on the Continent—at the price of rendering Western Europe dangerously, perhaps fatefully vulnerable to Soviet power. . . . When you tally up the total score of the two wars, in terms of their ostensible objective, you find that if there has been any gain at all, it is pretty hard to discern. Does this not mean that something is terribly wrong here?"[2]

We are today repeating the same folly, because the essential conditions that determine a peace settlement have not changed sufficiently since World War I. These conditions are, broadly speaking the following: First, the social sciences must have advanced to a point where they could reliably specify the requirements of an enduring peace. Second, social scientists must have attained such public respect that their voices would be influential at the peace table. The second condition, especially, is far from realization.

Some will answer that what we need is not primarily more knowledge, but the "will" to apply what we already know. But this "will" is itself a product of certain conditions. Why don't we apply what we do know? That is a question which applies to physical as well as to social knowledge. The remedies for certain diseases are sometimes known for decades or generations before the population as a whole can be induced to adopt the cures. Why is this? The reasons are numerous and intricate. Indeed, this question is itself a legitimate subject for research.

[2] George F. Kennan, *American Diplomacy, 1900–1950* (Chicago: University of Chicago Press, 1951), pp. 55–56.

Here again, there will be those who think they already know the answers. The most common of these answers is that education is the remedy. Just tell everyone of the scientific remedies. Tell everyone, also, that he is silly to have the habits, the inhibitions, and the superstitions which deter him from adopting the latest reliable scientific findings. The concrete details by which this ostensibly quick and easy program is to be achieved are themselves somewhat a mystery, not to say a superstition, for education is not a lever mounted and operated outside of the society which it aims to elevate. Education is a process grounded in, supported and manned by that same backward, silly, and superstitious society which it seeks to transform. Why don't we apply what we already know? The fact is, we have about such questions very little positive knowledge—verifiable generalizations—of the type that we have in other sciences and by virtue of which they perform their marvels.

In any event, whether education will help to solve our social problems depends upon the validity of what we teach. There is no doubt that much of what we now teach on social subjects is worse than useless, because it consists merely of transmitting the errors, prejudices, and speculations of bygone generations. Unless knowledge is constantly tested and replenished through scientific research, education may be an enemy rather than an aid to social amelioration.

Another popular variant of the doctrine that we already know all we need to know about social problems is the feeling that scientific solutions, even if they existed, could not be applied. There are always factions, it is pointed out, who profit, or think they profit, from the very maladjustments that constitute social problems, including war, economic depressions, and discrimination against minorities. Such a state of affairs is by no means unique to the social sciences. The application of every advance of the physical and biological sciences has been similarly opposed by those who had vested interests in the ignorance, the superstition, or the fraudulent reform movements antedating the development of scientific solutions. These vested interests were defeated in the end only by the demonstrable superiority of the solutions of science.

It is not true, therefore, that scientific solutions of social problems face a peculiar situation in that large numbers of people do not want such solutions and would be under no compulsion to accept them. Scientific solutions, in the long run, carry with them their own compulsions for acceptance. The demonstrated superiority of scientific methods has been, in the last analysis, the major reason why they have triumphed over the vested interests that opposed science. Also, once scientific criteria are accepted in a community as the final arbiters, no one challenges their decisiveness. If differences of opinion and interest arise as to the purity of water or whether a person is afflicted with a contagious disease, we accept the decision of scientists, although it may involve individual financial loss and the forcible incarceration of a free citizen.

The objections to social science reviewed above, although unsound, nevertheless appear to most people fairly conclusive against the possibility or desirability of looking to science for light on human relations. As a result, the solution of social problems is sought elsewhere. Considerable numbers still lift their hands for help to the sky. This may take the form either of looking for a saviour or in reading the stars. Twenty-five thousand people are reported by the newspapers (November, 1945) to have assembled in New York to witness the appearance of the Holy Virgin. Fortune-tellers—readers of tea leaves, palms, cards, handwriting, and astrologist-numerologists—are defrauding the people of the most "advanced" "scientific" country of hundreds of millions of dollars annually. A good deal of legislation has been enacted in an attempt to protect people against their own gullibility. But, as one popular magazine puts it, "whereas prophecy is illegal in almost every state, there's no law against giving advice on human problems"!

Others, who feel very superior to the sky-gazers, but who closely resemble them, believe instead in the appearance of The Great Man—The Great Leader. Recent history has afforded striking examples, in more than one country, of the congeniality as well as the absurdity of this belief. A variant of this view is the doctrine that, whether or not the voice of the people actually is the voice of God, *vox populi* is in any case just as good or better. "Fifty million people can't be wrong" is regarded as a conclusive refutation of any scientific finding about social or economic questions. Indeed, the late President Roosevelt reported that he looked hopefully to every mail for possible solutions to the major problems of the day. We would regard it as absurd to decide the efficacy of sulpha drugs by popular vote. Yet we feel very proper, democratic, and enlightened when we decide technical social questions by precisely that method.

The question of *whether we want to employ* either sulpha drugs or particular social remedies after their nature and efficacy are known may properly be settled by popular vote. Failure to grasp this distinction between the proper place of scientific authority and popular will in human relations is today a major source of difficulty, especially in countries that try to maintain democratic institutions. The confusion is also largely responsible for the failure to develop adequate scientific guidance in social as in physical matters. In the meantime, we may as well recognize the superficiality of the democratic variant of the fundamentally authoritarian mentality and emotional craving of our time. What sort of a social order do vast numbers of men, even in "democratic" countries, dream up when they give free play to their aspirations and longings? Is it an order in which truth is determined by the laborious methods of science and applied by democratic methods? It is nothing of the sort. The social structure of heaven, according to all authorities, is an absolute, albeit a benevolent, dictatorship.

We need to beware against making fetishes of words like Fascism, Communism, and Democracy. We would do better to concern ourselves with concrete questions of how and to what extent a given proposal, regardless of what it is called, is likely to minister to the broad satisfactions that nearly all men seek. This is where the scientist rather than the idealist and the moralist is indispensable.

The idealist is fond of pointing out that "where there is no vision, the people perish." Science certainly has no quarrel with this adage. All scientific activity is a quest in pursuit of a vision, namely, the hypothesis which gives direction and meaning to scientific enterprise. No one knows better than the scientist, therefore, the importance of vision in human activity. But the scientist also knows that a great deal depends upon the nature of the visions he elects to pursue. People who do not show a certain discrimination in the visions which they see and follow are, in fact, carted off to asylums to protect them and their fellow citizens from untimely and unseemly ends. In short, where there is irresponsible and fantastic "vision," the people are also likely to perish.

Perhaps it would be well for the great exponents of "vision" and "ideals" in human affairs to give more emphasis to the practical side of the matter than is customary. To be sure, the idea is implicit in most discussions of the subject. That is, when people advocate vision and ideals, they usually mean desirable, attainable goals. Nevertheless, a more careful scrutiny of social ideals and visions is perhaps one of the crying needs of our time for pretty much the same reason that it has been found necessary to restrict or supervise the applied idealism of advertisers, patent medicine venders, faith healers, and stock salesmen. It is doubtless a shocking thought that the visions for which men sometimes die and the ideals for which leaders are venerated should sometimes be indistinguishable from flagrant fraud and colossal ignorance. A sober survey of the history of idealism will nevertheless reveal such to be the case.

We must here distinguish sharply between those idealists who have specifically declared that the kingdoms they envision are definitely not of this world and those who equally specifically declare their programs to be for all men everywhere and in our time. Certainly no one objects to the visions of seers and sages regarding the ultimate destiny and complete happiness of man. Nor does science contemplate depriving man of romantic literature or of all the satisfactions which indubitably derive from fable, myth, and story. There is no question that these imaginative realms greatly enrich life and influence conduct.

For this reason, also, the somewhat indiscriminate deprecation of escapism is quite unwarranted. All that needs to be deplored is the inability to distinguish between fact and fable, the practical and the fantastic. Escape from the cares of the day has always been sought, has been anticipated with joy, and has refreshed the body and the mind. Different people will find this escape in different ways. Some find it in mathematics and philosophy, some in religion, some in music, some in poetry, and some at the movies in rocket trips to distant planets. I have no objection even to the last mentioned, provided the beholder does not try to take off in his car on the way home. This is precisely the important qualification that must be insisted upon regarding all historic lore, imaginative literature, visions, and ideals. Unbridled social idealism, unbalanced by scientific criteria as to possibilities and cost, is a social liability and in effect a type of fraud on the body politic.

Social scientists are at a considerable disadvantage today in competition with professional social idealists. Honest physicians were once at a similar disadvantage in relation to the snake oil practitioners. The latter promised quick and painless cures—no painstaking diagnosis or complicated course of treatment over a period of time was required. The offerings of social scientists today seem weak, uncertain, painful, and costly as compared with the proposals of the professional idealists whose lyrical sales talk clogs the airways, fills the press, and thunders from the platform. Science is not without its own ideals. But it has the decency to stipulate also the degree of probability of their attainment in given circumstances and periods of time.

The time will perhaps come when the irresponsible idealist will be regarded no more highly than other venders of bogus gold bricks. Today idealists go up and down the land with offerings that are as fraudulent as any patent medicine or snake oil that was ever retailed. Yet these are the respectable leaders of the day. It is true that the snake oil vender usually knew he was a fake, whereas many social idealists believe what they say. This, however, does not affect the truth or the frequently disastrous results of their respective offerings. Hitler's reported advice with respect to lying, namely, that a big lie is believed more readily than a small one, seems to hold also for a certain type of social idealism. A local relief program that aggravates the problem it is designed to cure is likely to arouse prompt and serious criticism. A world-wide "foreign-aid" policy, however, receives considerable support (both for humanitarian and political reasons) although it is easily demonstrable that every increment of relief (without a corresponding adjustment of birth rates and/or death rates) results in more hungry people than there were before the aid program was adopted. (See pp. 46–47.)

Those who demand the specifications of these grandiose projects and a look at the balance sheet of estimated costs and probabilities of success are castigated as cynics and ignoble characters. Nevertheless, a decent respect for certain facts and principles now known by social scientists would greatly modify the nature of projects embarked upon today by the statesmen of all nations, to the great advantage of their constituents in every land.

The function and power of ideals in human affairs has nowhere been questioned. Indeed, the present essay is itself idealistic. But scientists do not advance counsels of perfection nor advocate unattainable goals. Scientific validity is itself a matter of degree and, in new fields especially, it is sometimes difficult to distinguish the scientist from the quack. The fact remains that we have adopted in some fields fairly reliable criteria to distinguish science from nonscience. The social sciences could greatly assist in perfecting these criteria. There is, unfortunately, in science, as elsewhere, a type of idealism closely resembling that criticized above. It demands that social scientists present here and now formulations comparable in precision and predictive power to those of the advanced sciences. This we cannot be expected to do at present. We have rather taken the position that in science and idealism half a loaf is better than loafing, and that the road of science is a long, rough road. Ideals we must have. Let us distinguish between ideals and illusions.

III

A Common Outlook

The preference for prescientific thoughtways regarding our social relations has given rise to a fundamental and disastrous cleavage in our culture, for our social predicaments are what they are precisely because we have adopted and applied science so extensively in our relations with the physical world. "This cleavage," as John Dewey has said, "marks every phase and aspect of modern life: religious, economic, political, legal, and even artistic. . . It is for this reason," he goes on to say, "that it is here affirmed that the basic problem of present culture and associated living is that of effecting integration where division now exists. The problem cannot be solved apart from a unified method of attack and procedure." [3]

This unified method of attack must be that of modern natural science applied fully to human society, including man's thoughts, feelings, and "spiritual" characteristics.

I am aware of the fundamental nature of the revolution in methods of thinking which this proposal involves. I am aware that the scientific mode of thought is very recent in human history, that it is practiced by only a very small percentage of our own generation, and that it is uncongenial to large numbers of otherwise admirable people. Most people, as we have seen, prefer more immediate and miraculous solutions. They prefer to believe that a new political party, a "new deal," a new man in office, or especially a world organization is the solution. I am not

[3] More recently C. P. Snow has made the same point in characterizing contemporary Western societies as split into "two cultures," the scientific and the literary. I have for many years used the terms "legalistic-moralistic" versus the scientific to designate essentially the same phenomenon. See C. P. Snow, *The Two Cultures and the Scientific Revolution* (Cambridge, England: Cambridge University Press, 1959).

opposed to any of these, nor do I advocate that these interests be neglected. But I am pointing out that they cannot of themselves avail, unless an underlying body of knowledge is developed and employed which will be comparable to that which underlies engineering, navigation, and medicine. The ablest and most devoted leader cannot lead without charts and instruments relevant to the task at hand. These charts and instruments for navigating the social seas are as yet rudimentary and unreliable, even when they are employed. The consequences are on the front page of every newspaper.

I know that the method I propose is scoffed at in some quarters on the ground that, while it may be the solution for the long run, life is a short run. Whose life? Human life in its collective aspect stretches backward at least a million years and may reach much farther into the future. Throughout the ages it has been regarded as the mark of the intelligent and the civilized man that he has the power to feel and plan beyond the immediate situation. In any case, what choice have we? The proximate and immediate solutions which all desire are themselves dependent upon the development and use, in however modest degree, of scientific knowledge. Whether we look at it from the long or the short perspective, certain social problems have arisen as a result of science, and in our struggle with these problems science alone can save.

I have not said that science is all of life or of knowledge. The current controversy about science *versus* the humanities and the arts is, as we shall see, quite absurd. The assumption seems to be that the advancement of science and especially of social science must be at the *expense* of the other intellectual, artistic, and religious pursuits of man. This is a preposterous assumption. It implies that human knowledge is a finite quantity so that any increase in one department must be at the expense of some other. Actually, the advancement of science can only liberate, stimulate, and advance also the arts. As for the humanities and classics, regarding which there is much worry in certain quarters today, my proposal is to humanize science rather than to dehumanize humanity. The classics which truly fulfill the definition of classics are in no danger. It is, however, the privilege of each generation to define and choose its own classics. By broadest definition, they represent simply the highest excellence in their respective fields. What possible reason is there to assume that man will ever abandon or discard those things which he still finds of highest excellence?

The outstanding characteristic of our time, then, is confusion and contradiction in our thinking and methods of approach to human affairs. In our relations with the physical world, we have developed and employed a unified method which has been conspicuously successful in reaching demonstrably valid conclusions. We have failed signally to develop a similarly unified method of attack upon social problems. As a result, we find ourselves in the midst of strange and shattering frustrations. While there is great diversity of opinion regarding remedies, there is remarkable unanimity regarding the symptoms. They have become trite with repetition. Preachers, teachers, social workers, journalists, scientists, and statesmen call attention to the fact that, even in the most favored nation, millions know want in the midst of abundance. Other millions die from preventable disease. Finally, nations pursue each other throughout the forests and fields and jungles and over the seven seas with the most deadly weapons of destruction, loudly protesting that they are fighting for survival and in defense of all that makes life worth-while. Is it strange that under these circumstances many people should ask the question: Cannot science also do something in this desperate situation? We turn now to a consideration of that question.

CHAPTER II

Can Science Solve Social Problems?

I

Alleged Obstacles to Social Science

TRADITIONAL THOUGHTWAYS such as we have already reviewed are perhaps chiefly responsible for the failure thus far to take the social sciences seriously. It would be unfair to imply, however, that all skepticism regarding social science is of this character. Many thoughtful people, including scientists of distinction and unquestioned competence in their own fields, genuinely feel that there are certain differences between the subject matter of the physical and the social sciences which preclude the applicability of the same general methods to both. I do not refer here to the lapses of the great and near-great in physical science, literature, and art who frequently make childish pronouncements on the social order. I have in mind rather the serious student who has given some consideration to the matter.

Some eminent scientists,[1] justifiably disturbed over the careless way the words "scientific" and "scientific method" are currently used have expressed doubt regarding the propriety or advisability of speaking about *the* "scientific method." They take the position that the methods of the different sciences and of different scientists are of great variety and even to a high degree peculiar to the individual scientist. Without contradicting the position of these men

[1] E.g., J. H. Hildebrand, *Science in the Making* (New York: Columbia University Press, 1957), pp. 8–9. Also, J. B. Conant, *On Understanding Science* (New York: New American Library of World Literature, Mentor Edition, 1951), pp. 20–26.

in what they have to say regarding the shortcomings of much work in all the sciences and especially in the social sciences, it seems to me to be an overstatement to say or to imply that there are *no* rules of logic as well as *no* rules of procedure that characterize distinctively the behavior of scientists as scientists. If by "scientific method" or "methods" is meant the concrete procedures and paraphernalia of the different sciences, there can be no question regarding the variety of scientific procedures. After all, microscopes are employed in some sciences and telescopes in others. But surely there are generally valid principles and *methods* of sampling, correlation, probability computation, and logical inference to which nearly all scientists subscribe. Surely every scientist has in mind *some criteria* on the basis of which he undertakes to characterize both his own work and that of others as "scientific" or nonscientific. *These criteria* are what I have in mind when I use the phrase "scientific method" in this book. (An attempt to state them more generally will be found on page 78.) In the meantime, I shall deal with some of the more specific objections which have been raised against the possibility of social scientists measuring up to criteria which justify their classification as scientists.

For example, one distinguished scientist has urged that a basic difference between the physical and the social sciences is that in the latter "the investigator is inside instead of outside his material."[2] This is supposed to be so self-evident as to require no analysis. It turns out on examination to be little more than a figure of speech designed to call attention to the danger of biased observations

[2] Julian Huxley, "Science, Natural and Social," *Scientific Monthly,* Vol. 57, January, 1940. For a more comprehensive discussion of this and other alleged obstacles, see G. Lundberg, "Alleged Obstacles to Social Science," *Scientific Monthly,* Vol. 70, May, 1950, pp. 299–305, and "The Senate Ponders Social Science," *Scientific Monthly,* Vol. 64, May, 1947.

and interpretation—a danger which is present in all science and which can, in any case, be circumvented or reduced only through the use of scientific instruments and methods of procedure which are part of scientific training in *any* field.

When an anthropologist goes to a remote preliterate tribe to study its social behavior, why is he any more inside his material than when he studies a colony of anthropoid apes, beavers, ants, the white rats in the laboratory, an ecological distribution of plants or, for that matter, the weather, the tides, or the solar system? When a biologist studies his own body or takes his own temperature he is presumably very much inside or part of the phenomena he studies. At what point exactly in this series does the mysterious transition from outside to inside of one's material take place? Nor is it necessary to go to a distant land and study savages in order to make the point. I can give as objective and verifiable a report on some of the social events that take place in the community where I now live as I can on the meteorological events that take place there. Both involve problems of observing and reporting accurately the events which I am studying.

In doing this, we need to use instruments as far as possible to sharpen our observation, to check it, and to report it accurately. These instruments and skills do not exist ready-made in any field. They have to be invented. They may be quite elementary as yet in much social investigation, consisting of little more than a pencil, a schedule, a standardized test, or the recording of an interview. But we also have at our command the movie camera with sound equipment with which social behavior can be observed in its cruder aspects with the same accuracy as any physical behavior is observed. When I use these instruments, I am no more "inside" my material than when I photograph an eclipse.

The invention of units and instruments with which to systematize observation is part of the scientific task in all fields. (Neither calories nor calorimeters came ready-made in the phenomena of physics.) They have to be invented to apply to the behavior in question, just as units of income or standard of living and scales for measuring them have to be invented. I am not making light of the difficulties involved in inventing either such units or appropriate instruments of scientific observation. Nor should we minimize the problems of interpreting the data which we observe. But here again we have at our dipsosal the same rules of logic, statistics, and scientific method that we apply to observations of physical events.

Most people's acquaintance with science has involved laboratories and controlled experiments. Indeed, the word science probably conjures up to most people the image of a man in a white coat looking critically at a test tube. Accordingly, another insuperable obstacle to social science is usually urged. How can a piece of society be put in a test tube?

The importance of laboratory experimentation in the advancement of certain sciences cannot be denied. But the matter of laboratory control varies greatly with different sciences. The solar system has never been brought into any laboratory. Astronomical laboratories do contain very ingenious symbolic and mechanical representations of the astronomical aspects of that system and remarkable instruments for observing it. These every science must unquestionably develop. Beyond this, the questiion of laboratory conditions becomes one of convenience and mechanical ingenuity. Statistical devices which permit the observation of two or more variables while the others are held constant, in the sense that their influence is measured and allowed for, are already in common use. In any event, actual experimentation in the social sciences is not impossible. The interested reader will find it worth his while to examine the literature on this subject.[3]

II

What about Motives?

Another fatal obstacle to a full-fledged natural science of human social phenomena is alleged to be the presence in the latter of a unique and mysterious something called *motives.* When we start investigating human motives, the investigators' own motives become involved, we are told.

What is meant by motives? The word is used to designate those circumstances to which we find it "reasonable" to attribute an occurrence. The motive which we impute to an act is accordingly entirely relative to the frame of reference we adopt and accept as reasonable. The same event may be attributed to economic motives, the Oedipus complex, or the conjunction of the planets, according to whether one is an economic determinist, a Freudian, or an astrologer. To a scientist, the motives of a stone rolling downhill or of a boy murdering his father are simply the full set of circumstances resulting in either event. These conditions are equally subject to scientific investigation in both cases.

The circumstances, of course, will be of a very different type in the two cases. They will involve, in the case of the boy, "mental" and emotional factors of great subtlety ("complexity," "intangibility," "subjectivity"). But unless we are prepared to contend that for such phenomena other methods than those of science are more reliable, we have in fact no choice but to develop and improve scientific techniques appropriate to such investigation. Already our progress in this direction is not inconsiderable. Psychiatric and criminological techniques, including such devices as

[3] For a summary and a critical discussion, see E. Greenwood, *Experimental Sociology* (New York: King's Crown Press, 1945).

lie-detectors and brain-wave machines, strongly suggest that human motives are anything but immune to scientific investigation.

As for the scientist's own motives becoming involved, the only motive of a properly trained scientist in the face of a scientific problem is to solve the problem according to criteria specified by science. In short, the motive of the sociologist and physicist are exactly the same in the face of a scientific problem. *The motive is to find an answer that meets the requirements of a scientific answer.* The fact that the social scientist has always been a part of a social structure is no more a handicap to its objective study than is the fact that he is also part of the physical universe which he studies. Error, corruption, and bias, conscious and unconscious, are constant dangers inherent in *all* observation, both physical and social. Proper scientific training teaches the ethics of science as well as technical skill and the use of corrective instruments to reduce to a minimum the errors that beset our unchecked senses in every field.

Those who are oppressed with the feeling that the methods of science are inadequate in the determination of motives would do well to consider the methods by which motives are, in fact, determined today. One common method is to assemble twelve good men and true, farmers, bookkeepers, salesmen—anybody that comes to hand—subject them to a mass of evidence regarding the circumstances of a certain event, after which the judge charges them to determine whether or not "the motive" of one of the principals in the affair was fraudulent, felonious, malicious, etc. And how do these worthy citizens proceed with the assignment? They draw on their own lives and experience, on what their random and unsystematic observation of human behavior in the past has taught them, plus what they may have learned from folklore, the Bible, and perhaps books on history and psychoanalysis. Against this background they lay the testimony, and on this basis they decide "the" motive. The procedure is far from perfect, as we know, and it is not claiming too much to assert that scientists properly trained for this kind of work would make fewer mistakes. But we think highly enough of even the ability of the lay jury to determine human motivation to make their decision the basis of life or death for free citizens. The very crudeness of present methods of determining human motives is a good reason for turning to science in this important problem.

III

Instruments in Science

In view of the prevalence of the hopelessly biased social and political proposals that are currently offered in the name of science, it is not surprising that many people should conclude that the preference of the researcher influences all conclusions in social research. Have any of the corrective instruments mentioned above actually been produced?

There are scores of instruments in use today in the social sciences that detect, reduce, or measure the bias of our senses and the prejudice of different observers. Have we any examples of sociological conclusions uninfluenced by the author's likes, dislikes, and group affiliations? Let each one answer for himself. Does a *scientific* public opinion poll predict the election results irrespective of the sentiments of the poll-taker? Does the prediction of the number of marriages, divorces, school enrollment, or success in college vary according to the marital or educational status or hopes of the predicter? These personal characteristics and wishes in no way influence the predictions of a properly trained social scientist. Or consider the Census Bureau's facts and generalizations regarding trends in our population, labor force, and income. Are these conclusions Communist, Capitalist, or Fascist? The question makes no more sense than to ask whether the law of gravity is Catholic, Protestant, or Pagan.

Any comprehensive review of the present status and achievements of the social sciences is obviously beyond the scope of the present discussion. Although I think it is unquestionably true that the social sciences have made, during the present century, more actual progress than in all preceding history, it would be absurd to pretend that this progress is, as yet, reflected to any great extent in our management of social affairs. Scientific information of a more or less reliable character is more widely diffused than ever before, but the scientific mode of thought has obviously made very little headway. Practically no one approaches the major social problems of the day in a spirit of disinterested scientific study. The idea that these problems are to be solved, if at all, by the use of instruments of precision in hands that do not shake with fear, with anger, or even with love[4] does not seem to have occurred even to many people who pass for social scientists. They have joined the journalist and the soapbox crusader in the hue and cry of the mob. Their supposedly scholarly works bristle with assessments of praise and blame, personalities and verbal exorcisms which have no place whatever in the scientific universe of discourse. Not only do these angry men pass in the public eye as great social scientists of the day, but they not infrequently presume to patronize honest scientists who stay with their proper tasks of building a science and the instruments by means of which any difficult problems are to be solved.

But behind this fog, this dust storm of books about the inside of various political movements, the private life and morals of its leaders, and the treatises on democracy, substantial work is going on. Men are patiently accumulat-

[4] Cf. R. L. Duffus, *Harper's Magazine,* December, 1934.

ing data about human behavior in a form which in the fullness of time will permit a type of generalization which has never before been possible. Some are engaged in the undramatic but fundamental work, basic to all science, of classifying the multitudes of human groups and behavior patterns as a first step toward the formulation of generalizations regarding them. Still others are pioneering in the construction of actuarial and other tables from which may be predicted not only the prevalence of births, deaths, marriages, and divorces, but also the probable relative degrees of happiness in marriage, the probable success or failure of probation and parole, and many other equally "human" eventualities. A wealth of valuable information and generalizations have already been developed about the social characteristics and behavior of populations, such as the distribution of wealth, occupations, mobility, intelligence, and the various conditions with which these characteristics vary. Important instruments have been invented in recent years for measuring opinion, status, social participation, and many phenomena of communication and interpersonal relations.

Indeed, the invention and perfection of instruments for the more accurate and precise observation and recording of social phenomena must be regarded as among the most important developments in the social sciences. It is easy to point to the flaws in these instruments as it was easy to point to the flaws in the early microscopes and telescopes. But, without these beginnings and the patient centuries of undramatic labor, sciences like bacteriology could not have appeared at all.

Finally, there are those, and they may be the most important of all, who are experimenting with and inventing new systems of symbolic representation of phenomena. New adaptations of mathematics by which otherwise hopeless complexities can be comprehended are quite fundamental but do not lend themselves to popular display. The work of Leibnitz, Faraday, and Hertz was not the popular science of their day. Yet is it by virtue of their strange calculations with strange symbols that men today fly and broadcast their speech around the earth. This should be remembered by "writers" and others who complain that social scientists are adopting "jargon" and "esoteric" symbols which go beyond the vocabulary of the current "best-seller."

If I deal primarily with these more obscure and undramatic labors of social scientists, it is because I regard them as more important in the long run than the conspicuous contemporary achievements which are common knowledge. I do not overlook or underestimate these more obvious and demonstrable achievements. The transition in our time to more humane treatment of children, the poor, and the unfortunate, by more enlightened education, social work, and penology must in large measure be attributed to the expanding sciences of psychology and sociology. I know, of course, that whenever a war or a depression occurs journalists and preachers point to the impotence of economists and poltical scientists either to predict or prevent these disasters. The fact is that the course of events following World War I, down to and including the present, was predicted with great accuracy by large numbers of social scientists. That nothing was done about it is not the special responsibility of social scientists. "Doing something about it" is the common responsibility of all members of a community, including scientists, and especially of those who specialize in mass education, mass leadership, and practical programs.

It is not my main purpose to review the past and present achievements of the social sciences. Some of these achievements will be considered in the next chapter. I am here concerned primarily with the probable future of the social sciences. Even if I should admit that social scientists are today merely chipping flint in the Stone Age of their science,

I do not see that we have any choice but to follow the rough road that other sciences have traveled. The attainment of comparable success may seem remote, and the labors involved may seem staggering. But is the prospect really unreasonably remote? Suppose that someone four hundred years ago had delivered an address on the future of the physical sciences and suppose that he had envisioned only a small fraction of their present achievements. What would have been the reaction of even a sophisticated audience to predictions of men flying and speaking across oceans, seeing undreamed-of worlds, both through microscopes and telescopes, and the almost incredible feats of modern engineering and surgery? Nothing I have suggested, I think, in the way of mature social science with comparable practical application seems as improbable as would the story of our prophetic physicist of four hundred or even one hundred years ago.

The time is passing when the solid achievements of social science and its future prospects can be dismissed with a faraway look, an ethereal smile, and the remark that unfortunately science neglects "*the* human factor." *The* human factor is apparently something as mysterious and inscrutable as the soul or other ectoplasmic manifestations. In any event, no one seems to be able to give any further light on the nature of *the* human factor. The assumption that there is any such single factor is itself gratuitous. What we do find on inquiry are a lot of human factors—all the loves, hates, jealousies, prejudices, fears, hopes, and aspirations of men. We know what we know about these factors through observation of human behavior and what more we need or want to know is to be learned in that way and in no other. There is nothing more mysterious about human factors than about other phenomena which have not yet become the subject of serious scientific study. Of course, the trades and professions which have a vested interest in obscurantism, mysteriousness and ignorance will oppose the advancement of science into this field as they have in other domains. But I suspect their efforts will not avail. Science has already so firm a hold on the imaginations of men that they will insist on invoking this powerful tool also in the explanation of the mysterious human factor. *The* human factor will then be found to be no factor at all, but merely a vague word designating a great variety of behavior which social scientists have hitherto been too lazy or ignorant to approach by the same methods that have clarified the factors in other phenomena of nature.

We have hitherto lacked boldness and an adequate vision of the true task of social science. Research in this field is today for the most part a quest for superfiicial remedies, for commercial guidance, and for historical and contemporary "human interest" stories. Everybody recognizes the importance of bookkeeping, census taking, studying the condition of the Negro population, and predicting the number of girdles that will be purchased in department stores a year from now. But there are types of research the immediate practical uses of which are not so obvious, yet which are essential to scientific development.

Shall we or shall we not assume that we can formulate laws of human behavior which are comparable to the laws of gravity, thermodynamics, and bacteriology? These latter laws do not of themselves create engineering wonders or cure disease. Nevertheless they constitute knowledge of a kind which is indispensable. The present argument is obviously handicapped in its most crucial respect, namely, its inability, in the space here available, to exhibit laws of social behavior comparable to the physical laws mentioned. Yet we have made considerable progress in this direction.

IV

Science and "Values"

Finally, we come to what is regarded by many people, including scientists, as the most fundamental difference of all between the physical and the social sciences. "To understand and describe a system involving values," says Huxley, "is impossible without some judgment of values." "Values," he goes on to say, "are deliberately excluded from the purview of natural science."

It would be difficult to find a better example of confused thinking than that offered by current discussions of "values" and their supposed incompatibility with science. A principal cause of the confusion is a semantic error which is extremely common in the social sciences. In this case, it consists in converting the verb "valuating," meaning any discriminatory or selective behavior, into a noun called "values." We then go hunting for the *things* denoted by this noun. But there are no such things. There are only the valuating *activities* we started with. What was said above about motives applies with equal force to values. They are clearly inferences from behavior. That is, we say a thing *has* value or *is* a value when people behave toward it so as to retain or increase their possession of it. It may be economic goods and services, political office, a mate, graduation, prestige, a clear conscience, or anything you please. Since valuations or values are empirically observable patterns of behavior, they may be studied as such, by the same general techniques we use to study other behavior.

As a matter of fact, everybody is more or less regularly engaged in such study of other people's values. It is quite essential to any kind of satisfactory living in any community. We try to find out as soon as possible what the values of our neighbors are. How do we find out? We observe their behavior, including their verbal behavior. We listen to what other people say about them, we notice what they spend their money for, how they vote, whether they go to church, and a hundred other things. On a more formal and scientific level, opinion polls on men and issues are taken to reflect the values of large groups. Economists, of course, have been studying for years certain kinds of evaluations of men through the medium of prices.

There appears to be no reason why values should not be studied as objectively as any other phenomena, for they are an inseparable part of behavior. The conditions under which certain values arise, i.e., the conditions under which certain kinds of valuating behavior take place, and the effects of "the existence of certain values" (as we say) in given situations are precisely what the social sciences must study and what they are studying. These values or valuating behaviors, like all other behavior, are to be observed, classified, interpreted, and generalized by the accepted techniques of scientific procedure.

Why, then, is the value problem considered unique and insurmountable in the social sciences?

The main reason seems to be that social scientists, like other people, often have strong feelings about religion, art, politics, and economics. That is, they have their likes and dislikes in these matters as they have in wine, women, and song. As a result of these preferences, both physical and social scientists frequently join other citizens to form pressure groups to advance the things they favor, including their own economic or professional advancement, Labor, Capital, Democracy, the True Church, or what not. To do so is the right of every citizen, and there is no canon of science or of civil law which requires scientists to abjure the rights which are enjoyed by all other members of a community.

The confusion about values seems to have arisen because both scientists and the public have frequently assumed that, when scientists engage in ordinary pressure-group activity, that activity somehow becomes science or scientific activity. This is a most mischievous fallacy. It is not surprising, perhaps, that the public should be confused on this point, because it may not always be clear when a scientist is expressing a scientific conclusion and when he is expressing a personal preference. But it is unpardonable for scientists themselves to be confused about what they know and say in their capacity as scientists and what they favor in religion, morals, and public policy. To pose as disinterested scientists announcing scientific conclusions when in fact they are merely expressing personal preferences is simple fraud, no matter how laudable or socially desirable may be the scientists' "motives" and objectives.

But is it possible for a person to play two or more distinct roles, such as scientist and citizen, without confusing the two? The answer is that it is being done every day. It is the most obvious commonplace that the actress who plays Juliet in the afternoon and Lady Macbeth at night does not allow her moral or other preference for one of these roles to influence her performance of the other. In any event, her competence is measured by her ability to play each role convincingly. During the same day she may also be expected to fulfill the roles of wife, mother, etc. Likewise, the chemist who vigorously campaigns against the use of certain gases in war obviously cannot allow that attitude to influence in the slightest degree the methods of producing or analyzing these gases. Science, as such, is non-moral. There is nothing in scientific work, as such, which dictates to what ends the products of science shall be used.

In short, it is not true that "to understand and describe a system involving values is impossible without some judgment of values." I can certainly report and understand the bald fact that a certain tribe kills its aged and eats them, without saying one word about the goodness or badness of that practice according to my own standards, or allowing these standards of mine to prevent me from giving an accurate report of the facts mentioned. The only value judgments which any properly trained scientist makes about his data are judgments regarding their relevance to his problem, the weight to be assigned to each aspect, and the general interpretation to be made of the observed events. These are problems which no scientist can escape, and they are not at all unique or insuperable in the social sciences.

Have scientists, then, no special function or obligation in determining the ends for which scientific knowledge is to be used? As scientists, *it is their business to determine reliably the immediate and remote costs and consequences of alternate possible courses of action, and to make these known to the public.* Scientists may then *in their capacity as citizens* join with others in advocating one alternative rather than another, as they prefer.

There appears to be a good deal of misunderstanding of my position as stated in the preceding paragraph. Many earnest scientists as well as literary men seem to feel that my distinction between the scientist, as scientist, and the scientist, as citizen, is somehow invalid or misleading. Thus C. P. Snow points out that the scientist, by virtue of possessing certain knowledge, thereby becomes saddled with "a direct and personal responsibility. It is not enough to say that scientists have a responsibility as citizens. They have a much greater one than that, [?] and one different in kind. [?] For scientists have a moral imperative to say what they know."[5] (Brackets mine)

I certainly agree that scientists have this moral imperative —*both* in their capacity as scientists and in their capacity as citizens. It is one of the requirements of science for the sci-

[5] C. P. Snow, "The Moral Un-Neutrality of Science," Address to the 127th Meeting of the A.A.A.S., New York, 1960, printed in *Science*, January 27, 1961, pp. 255–61.

entist to publish his results and how they were reached. The implication that nonscientists do not have this responsibility, this moral imperative, seems to me untenable. *Both* the scientist and the nonscientist *in their capacity as citizens* have a "moral imperative" "to say what they know" if they happen to know that a time bomb is ticking away under the city hall or that the water supply is being subtly poisoned. It seems absurd to me to say that the scientist also *in his capacity as scientist* has a moral imperative to prevent the city hall from being blown up. The question of whether or not to blow up the city hall is not a scientific question, and it is not the type of question which any science, and least of all, physics, can answer. To say what one knows may involve costs and consequences of the most serious kind, but these are consequences to which both scientist and nonscientist are subject. Fortunately, as Snow has pointed out, scientists perhaps have at least as much moral stamina as nonscientists in such matters. Thus we escape the greater disasters to which we might otherwise be exposed.

An illustration may clarify the point. In the report of the hearings of a U. S. Senate Committee on the bill establishing the National Science Foundation there occurs this item:

Senator Fulbright: I asked an able scientist yesterday if he would define social science. I had been worrying about that. He said in his definition, "In the first place, I would not call it science. What is commonly called social science is one individual or group of individuals telling another group how they should live." (8164) [6]

As is well known, this is a very generally accepted view and, unfortunately, perhaps in large part a warranted view. My point is that it is not *necessarily* true, and that one aim of all social scientists should be to make it always *false*. If so,

[6] *Hearings Before a Subcommittee of the Committee on Military Affairs,* U. S. Senate, 79th Congress, first session, Parts I–V. Number in parentheses following quotation refers to page in the *Congressional Record.*

social scientists should first agree that the sole function of scientific work is *to grind out and publish systematically related and significant "if . . . then" propositions which are demonstrably probable to a certain degree under given circumstances.* Under this definition "to say what one knows," i.e., *to publish one's findings* is certainly a clear imperative. Any advocacy or attempt on the part of a scientists in his capacity of *scientist* to specify what scientific knowledge is to be used for is equally clearly outside the scientific sphere as defined. *In his capacity as citizen,* the scientist may of course advocate anything his heart desires with all the passion and resources at his command. If we do not rigorously insist on this distinction between scientific knowledge as contrasted with all other forms of knowledge whatsoever, we invite the corruption of science by injecting into it the biases of human preferences, tastes, and values which is precisely the charge today laid at the door of the social sciences.

Many sincere and thoughtful people, while agreeing for the most part with the position advanced above, nevertheless hesitate at some of its implications. The question takes this form: Presumably the ideal scientist, in his role as scientist, has allegiance to only one belief, namely, that the presently developing methods of modern scientific inquiry are more likely than any other methods to yield useful warranted assertions for the guidance of men in society. Does it not follow that such a scientist, in his role as such, should advocate whatever form of government proves most favorable and least obstructive or inhibitive to such inquiry?

There can be no doubt, it seems to me, that if my definition of science and scientific method is accepted, my answer to the question must be an unqualified negative. I agree that a scientist, in his capacity as such, might find that *if the criteria* of governments more and less favorable to the advancement of science are specified, it is possible for the scientist strictly in his capacity as scientist to say that Government A is preferable to Government B, because the criteria now are the "if" in an "if . . . then" proposition. The mores of science require him to publish this finding, *including the criteria.* I know of no canon or science which requires *or permits* him *as a scientist* to *advocate* the form of government which according to the stipulated criteria is "best" for the advancement of science. But I think it is both his right and his duty *as a citizen* not only to publish his finding (which is required also by the mores of science) but to advocate that form of government. The crucial point is that the scientific conclusions *depend upon the criteria.* And *the criteria* were not arrived at by scientific procedures. They were *taken,* as they always must be—they are the IF in all scientific conclusions. And all scientific conclusions involve and include the *"if." All* scientific generalizations hold only under stipulated conditions. *No* scientific laws *in any science* hold *except* under stipulated conditions.

To the extent that their reputation and prestige is great, and to the extent that their tastes are shared by the masses of men, scientists will, of course, be influential in causing others to accept the goals the scientists recommend. In this sense, social science will doubtless become, as physical science already is, an important influence in determining the wants of men. That is, as a result of scientific knowledge, men will not want impossible or mutually exclusive things. They will not seek to increase foreign trade and at the same time establish more comprehensive and higher tariffs. They will not seek to reduce crime but at the same time maintain a crime-promoting penal system. They will not destroy the productive power of a nation and still expect it to be peaceful, prosperous, and democratic. They will not expect a world organization to be conjured into existence by semantically deranged "statesmen," before the necessary preceding integration of the constituent units has been achieved.

The development of the social sciences and the diffusion of scientific knowledge will doubtless greatly influence in the above ways the wants, wishes, and choices of men. But there is still an important difference between a statement of fact and the dictation of conduct. It is one thing for a physician to tell a patient: "Unless you undergo this operation, which will cost so much in time, money, and pain, you will probably die in one month." It is another matter to say: "Science, for which I am an accredited spokesman, says you shall undergo this operation." Any scientist who pretends that science authorizes him to make the latter statement is a fraud and a menace. Dictation of this type has not accompanied the rise of physical science and it need not result from the full maturity of the social sciences. This needs to be kept in mind especially in these days of much worry about brain trusts and whether, with the development of atomic fission, scientists must become a priestly class dictating all public policy.[7]

The misunderstanding regarding the relation of scientists to practical affairs is so widespread and mischievous as to warrant further emphasis. The *application* of scientific knowledge obviously involves value judgments of some sort. This problem is equally present in the other sciences. After we know how to produce dynamite and what it will do, there remains the question: Shall we drop it from airplanes to destroy cathedrals and cities, or shall we use it to build roads through the mountains? After we know the effects of certain drugs and gases, the question still remains: Shall we use them to alleviate pain and prevent disease, or shall we use them to destroy helpless and harmless populations? There is certainly nothing in the well-developed sciences of chemistry or physics which answers these questions. Neither is it the business of the social sciences to answer (except

[7] For further elaboration of this whole subject, see G. Lundberg: "Science, Scientists, and Values," *Social Forces*, May, 1952, pp. 373–79.

conditionally, as we have seen) the question of what form of government we should have, what our treatment of other races should be, whether we should tolerate or persecute certain religious groups, whether and to what degree civil liberties should be maintained, and a multitude of other questions which agitate us. What, then, are social scientists for and what should they be able to do?

Broadly speaking, it is the business of social scientists to be able to predict with high probability the social weather, just as meteorologists predict sunshine and storm. More specifically, social scientists should be able to say what is likely to happen socially under stated conditions. A competent economist or political scientist should be able to devise, for example, a tax program for a given country which will yield with high probability a certain revenue and which will fall in whatever desired degrees upon each of the income groups of the area concerned. Social scientists should be able to state also what will be the effect of the application of this program upon income, investments, consumption, production, and the outcome of the next election. Having devised such a tax program and clearly specified what it will do, it is not the business of the social scientists any more than it is the business of any other citizens to secure the adoption or defeat of such a program. In the same way, competent sociologists, educators, or psychologists should be able to advise a parent as to the most convenient way of converting a son into an Al Capone or into an approved citizen, according to what is desired.

My point is that no science tells us *what to do* with the knowledge that constitutes the science. Science only provides a car and a chauffeur for us. It does not directly, as science, tell us where to drive. The car and the chauffeur will take us into the ditch, over the precipice, against a stone wall, or into the highlands of age-long human aspirations with equal efficiency. If we agree as to where we want to go and tell the driver our goal, he should be able to take us there by any one of a number of possible routes the costs and conditions of each of which the scientist should be able to explain to us. When these alternatives have been made clear, it is also a proper function of the scientist to devise the quickest and most reliable instrument for detecting the wishes of his passengers. But, except in his capacity as one of the passengers, the scientist who serves as navigator and chauffeur has no scientific privilege or duty to tell the rest of the passengers what they *should* want. There is nothing in either physical or social science which answers this question. Confusion on this point is, I think, the main reason for the common delusion that the social sciences, at least, must make value judgments of this kind.

But it does follow, as we have seen, that science, by virtue of its true function, as outlined above, may be of the utmost importance in helping people to decide intelligently what they want. We shall return to this subject in the concluding chapter. In the meantime, it may be noted that the broad general wants of people are perhaps everywhere highly uniform. They want, for example, a certain amount of physical and social security and some fun. It is disagreement over the means toward these ends, as represented by fantastic ideologies, that results in conflict and chaos. I have pointed out that, in proportion as a science is well developed, it can describe with accuracy *the consequences* of a variety of widely disparate programs of action. These consequences, if reliably predicted, are bound strongly to influence what people will want. But it remains a fact that science, in the sense of a predicter of consequences, is only *one* of the numerous influences that determine an individual's wants and his consequent behavior. Science and scientists are still the servants, not the masters, of mankind. Accordingly, those scientists who contend that they can scientifically determine not only the means but the ends of social policy should be exposed as scientific fakers as well as would-be dictators. Yet this is the very group which professes to be concerned about the undemocratic implications of the position I am here defending!

Finally, this view seems to some people to do away with what they call "the moral basis of society." Obviously, it does nothing of the sort. The question is not about the moral basis of society but about the social basis of morals. We merely advocate a scientific basis for morality. Presumably, all will agree that morals exist for man, not man for morals. Morals are those rules of conduct which man thinks have been to his advantage through the ages. Why should we then not all agree that we want the most authentic possible appraisal of that subject?

V

The Moral Neutrality of Science

There appears, then, to be no reason why the methods of science cannot solve social problems. Neither should we expect more from social than from physical science. As *science,* both physical and social sciences have a common function, namely, to answer scientific questions. These answers will always be of an impersonal, conditional type: "*If* the temperature falls to 32°F., *then* water (H_2O) will freeze." "*If* a certain type of tax is adopted, *then* certain types of industrial activity will decrease." Neither of these statements carries any implications as to whether or how the knowledge should be used. Far from being a weakness, this characteristic of scientific knowledge is its greatest strength. The wants of men will change with changing conditions through the ages. The value of scientific knowledge lies precisely in this impersonal, neutral, general validity for whatever purposes man desires to use it.

For this reason, those scientists and others who try to identify science with some particular social program, sect, or party must be regarded as the most dangerous enemies of science. They are more dangerous than its avowed enemies, because the defenders of "democratic," "communist," "religious," or "moral" science pose as defenders of science and carry on their agitation in the name of lofty social sentiments. That this group is confused rather than malicious is evident from their proposal that scientists should take an oath not to engage in scientific activity which may be "destructive" or contrary to "toleration," "justice," etc. The absurdity of the proposal is readily apparent, if we consider any actual scientific work. No scientist can foresee all the uses to which his work may be put, and in any event it is a commonplace that the *same* drug may be used to cure or to kill people. It may be granted that preposterous proposals of this kind are a temporary hysterical phenomenon superinduced by such dramatic developments as the atomic bomb. It may be granted that the agitators are motived by lofty social sentiments. Unfortunately, the same has been said for prominent proponents of the Inquisition.

The uses to which scientific or other knowledge is to be put have always been and will continue to be a legitimate concern of men. Science, as we have noted, can be valuable in helping men to decide that question. Our warning here has been directed against attempts to corrupt scientific methods and results by allowing them to be influenced by the temporary, provincial, ethnocentric preferences of particular scientists or pressure groups.[8]

[8] That such a warning is in order appears from a recent study by S. S. West. Only one-third of the fifty-seven researchers in six science departments of a major midwestern university took the view that the scientist, as scientist, is completely neutral in his concern for how the knowledge is applied. About one-fourth felt they could separate their roles as scientists and citizens, remaining neutral in the one but stating their position in the other. (Reported in "Scientists vs. the Ideology of Science," *The American Behavioral Scientist,* Vol. 4, March, 1961, p. 35.)

CHAPTER III

The Transition to Science in Human Relations

I

Consensus on Methods

IN THE preceding chapters I have expressed the view that the best hope for man in his present social predicament lies in a type of social science strictly comparable to the other natural sciences. We have reviewed some of the objections that have been urged both by physical and social scientists to this proposal. I am not under the illusion that my argument can be established conclusively in so brief a compass. Actually, of course, only time and future scientific development can finally demonstrate the validity of the position which I have outlined.

In the meantime, we are confronted with the necessity of proceeding on *some* hypothesis as to the way out of our difficulties. It is generally agreed, even by those who differ most radically as to the proper approach, that our first need is a unified, coherent theory on which to proceed. A society cannot achieve its adjustments by mutually incompatible or contradictory behavior, any more than can an individual organism. However we may differ on details and on ends, we must agree on certain broad means, certain principles of action toward whatever ends we do agree upon.[1]

In short, we all apparently agree with Comte's appraisal of the situation as he saw it a hundred years ago. Speaking

[1] See Chapter VI for a discussion of the possibility of such agreement among people of widely different beliefs and backgrounds.

of the theological, the metaphysical, and the positive scientific approaches, he said: "Any one of these might alone secure some sort of social order: but, while the three co-exist, it is impossible for us to understand one another upon any essential point whatever."

Of course there are some who find in our present predicament merely further evidence of the futility of the scientific approach in human affairs. They overlook the fact that, actually, science has as yet not been tried on social problems. Consequently, they advocate a return to theology, or "the" classics, either in their historic forms or in new versions in which the advocates of these approaches personally can play the role of major prophets. If I could see any chance of bringing about a return to theology or "the" classics, I might give it serious consideration, because any one unified approach might be better than two or more contradictory ones. But I see no such possibility in the long run. The commitments we have already made to science, chiefly in our technological culture, are of such character that we can neither go back nor stand still.

Our technological developments and our methods of communication have resulted in a fundamental interdependence which dominates our lives. This state of affairs requires, as we shall see, that we bring our social arrangements into line with this basic technological pattern, rather than vice versa. This basic technological pattern unquestionably rests upon natural science. On this ground, rather than on any assumption of absolute or intrinsic superiority of science as a philosophy of life, I think the following conclusion is inescapable: *In our time and for some centuries to come, for better or for worse, the sciences, physical and social, will be to an increasing degree the accepted point of reference with respect to which the validity (Truth) of all knowledge is gauged.*

If we accept this conclusion, then a number of questions

arise. (1) What are some examples of what the social sciences have done or might do in furthering sound and orderly adjustments in human relations? (2) What are some of the requirements and the costs of a transition to a social order in which science is the final court of appeal? (3) What would be the effect of such a transition upon democratic institutions?

The present chapter will deal with these questions.

II

What Can Be Done—Some Examples

What are some examples of types of work by social scientists that are of vast importance in managing human relations?

When we speak of *types* of work by social scientists, we are obviously announcing an undertaking so large as to prevent even a summary within the confines of this book. There are at least five well-recognized social sciences, and if we use the larger category of "behavioral science," the number rises to twelve or more. The social sciences are well-recognized in the sense that they are firmly established as departments in nearly all leading universities and colleges as well as in professional, industrial, and governmental circles. Over a hundred journals publish every year hundreds of research reports of studies large and small, designed to yield new knowledge or to test and refine previous conclusions and to predict behavior under stipulated conditions. We shall confine ourselves to a few illustrations [2] selected chiefly because they are individually of interest to more than one of the social sciences. Readers interested in more com-

[2] For a longer list, see Stuart Chase, *The Proper Study of Mankind* (New York: Harper & Bros., 1948), pp. 50–51. Chase lists some twenty "outstanding accomplishments" as of 1948.

prehensive accounts, including methodological details, will find a large literature readily available.[3]

For our present purpose we shall not here become involved in the question, touched in preceding chapters, of the degree of scientific refinement attained in the different sciences. My argument in previous chapters has been based in large part on what appears to me to be warranted anticipations regarding *future developments* of the social sciences. Here, and throughout the rest of the book, I shall rather take the view that, *even with their present shortcomings,* the social sciences must be taken seriously. The recent (1960) elevation of the Office of Social Sciences to full divisional status in the National Science Foundation [4] is an indication of this growing recognition.

The work of such agencies at the Census Bureau is known to all and is more or less taken for granted. Without the data and the analyses which it provides, the administration of public affairs would certainly dissolve in chaos and perhaps in civil war. It is equally certain that no international organization can function without an elaborate organization of this kind to provide the essential facts regarding people and their characteristics and activities. Perhaps the most permanently valuable contribution of the ill-fated League of Nations was its establishment of an international statistical bureau which managed to survive until taken over by the larger information agencies of the United Nations. The Office of Population Research at Princeton University has engaged in detailed studies of local and international population trends in various parts of the world and has

[3] See, for example, Gardner Lindzey, *Handbook of Social Psychology,* 2 Vols., (Reading, Mass.: Addison-Wesley, 1954). Also, R. K. Merton, L. Broom, L. S. Cottrell, Jr. (eds.), *Sociology Today* (New York: Basic Books Inc., 1959).

[4] The other divisions are: (1) Mathematical, Physical, and Engineering Sciences; (2) Biological and Medical Sciences; and (3) Scientific Personnel and Education.

predicted the future areas of population pressure. This knowledge is of the utmost practical importance in the administration of national and international organization of any kind. The Scripps Foundation, the Milbank Memorial Fund, and many others are engaged in similar or related work of a character that measures up very well to the standards of the physical sciences.

Social scientists have also been prominent in pointing out one of the most serious of the world's *problems,* namely, the problem of overpopulation. As a result of the drastic decline in the death rate resulting from the application of medical science, world population is increasing at an unprecedented rate. For example, although it took thousands of years for the human species to reach the number of one billion of living people (about 1830) it required only one century to add the second billion. It is now taking less than thirty-five years for the world population to add a third billion—probably before 1965. The United Nations' population experts estimate that it will take only fifteen years to add a fourth billion, and another ten years to add the fifth billion if present rates should continue. The idea that any expansion of the food supply could do more than temporarily alleviate the starvation of people under such rates of population increase is merely a confusion of wishful thinking with stern realities.

However, just as the application of science to health and sanitation has produced this situation, science has provided the means for its control. Further improvements in the latter are highly likely and imminent. The distinctively social problem of securing the widespread adoption of known methods of control involve a number of problems of a type not yet fully solved, but under extensive inquiry by social scientists. In the meantime we have an example of successful population control in the case of postwar Japan. We are not here concerned with these problems in themselves, but with the role of scientifically gathered and analyzed human social data in the prediction of future population, and the solution of a problem which some regard as more dangerous than nuclear war. Also in other ways, statistics of individual countries, and the data collected by the United Nations organization, are of fundamental importance to the work of many scientists engaged in a wide variety of particular projects. Human ecology, which cuts across the conventional boundaries of demography, geography, sociology, economics, political science (and perhaps others), has produced very impressive work both of applied and theoretical significance.

Reliable and objective knowledge of other peoples and cultures constitutes another field in which social scientists have made distinguished contributions. This knowledge has thrown a flood of light on our own civilization and permits the formulation and test of hypotheses regarding human behavior patterns in general. The Human Relations Area Files contain, systematically filed and indexed, virtually all present reliable knowledge regarding some two hundred cultures. To make a long story short, if a researcher happens to be interested in some subject as, for example, divorce, crime, education, law (and about a thousand other topics), in other cultures, he can go to one of the twenty or more libraries which subscribe to the File, and find all the known information on any or all of these subjects for each of about two hundred cultures. The information is neatly filed away in a separate drawer for each subject. Information which it might take years to locate as scattered in hundreds of books in a library can be secured in a few hours from the File. The importance of this kind of knowledge and its ready availability in facilitating our contacts with people of other lands and cultures became very evident during and after World War II.

We mentioned in the preceding chapter the importance of instruments and methods of observation and measurement in the social as well as in the physical sciences. Social scientists have produced revolutionary developments in this field in the last thirty years. Thousands of such instruments have been invented by means of which vocational aptitudes, success in college and other undertakings, and social behavior of great variety can be accurately measured and predicted. Instruments and scales for the measurement of attitudes have opened vast new fields for investigation.

Perhaps the best known, but by no means the only one, of these devices is the public opinion poll. We have in this technique an illustration of how a development in the social sciences may be as significant for the future of social organization as many physical inventions have been in our industrial development. The mechanisms by which the "public will" can make itself reliably felt in government and community action has always been in the foreground of political discussion. With the expansion of the areas in which public opinion must operate, many students of the problem have despaired of the capacity of the town meeting technique adequately to make operative the "public will." In the face of this situation, the scientific public opinion poll constitutes an instrument which cheaply and accurately permits us to learn the beliefs, the attitudes, and the wishes of the rank and file of the population. Public opinion polls are at present frequently thought of as interesting devices mainly for predicting the outcome of elections. They do permit such prediction, but this is a very minor aspect of their full possible importance. Polls were extensively used in the armed forces in World War II as a guide to the administration of the invaded areas, the return of the armed forces after the war, and in many other ways.

Public opinion polling may be a device through which can be resolved one of the principal impasses of our time, namely, the apparent irreconcilability of authoritarian control on the one hand and the "public will" on the other. It may be that through properly administered public opinion polls professionalized public officials can give us all the efficiency now claimed for authoritarian centralized administration and yet have that administration at all times subject to the dictates of a more delicate barometer of the peoples' wills than is provided by all the technologically obsolete paraphernalia of traditional democratic processes. In short, it is not impossible that as the advancing technology in the physical adjustments of our lives leads to a threatened breakdown of democracy, so an improved social research instrument may restore and even increase the dominance of the people's voice in the control of human society.

The time may come when the reliable polling of public opinion will be a science comparable to meteorology. Charts of all kinds of social weather, its movements and trends, whether it be anti-Semitism, anti-Negro sentiment, or mob-mindedness will be at the disposal of the administrators of the people's will in every land. A barometer of international tension has been designed to detect reliably and early the tensions that lead to war.[5] It is true that mere knowledge of these tensions does not necessarily operate to alleviate them. But it is also true that a reliable diagnosis of the tension and an understanding of the feelings and sentiments that underlie tensions is essential for an effective approach to the problem.

"Statesmen" will doubtless continue for some time to value their intuitions more highly than scientific prediction. Pious platitudes doubtless will continue to be heard about the "unpredictability" of human behavior. It remains a fact that social scientists predicted within a fraction of 1 per cent the actual voting behavior of sixty-eight million voters in the U.S.A. in the presidential election of 1960. The pollsters have been doing so regularly since 1936 with a maxi-

[5] S. C. Dodd, "A Barometer of International Security," *Public Opinion Quarterly,* Summer, 1945.

mum error of 6 per cent. Nor are such results limited to voting behaviors. The late Professor Stouffer of Harvard predicted, also within a fraction of 1 per cent, the number of discharged soldiers after World War II who would take advantage of the educational privileges of the G. I. Bill of Rights. Hundreds of other cases could be reported from a great variety of fields of human social behavior, including the vast areas of market research.

To those who constantly have their minds on quick and dramatic solutions to the world's troubles this type of research is likely to seem offensively trivial—a kind of fiddling while Rome burns. "Writers" are fond of referring contemptuously to basic scientific work as an "ivory tower" and as "lecturing on navigation while the ship sinks." Navigation today is what it is because some people were willing to study the *principles* of their subject while their individual ships went down, instead of rushing about with half-baked advice as to how to save ships that could not be saved, or were not worth saving anyway. As A. J. Carlson has recently said: "The failure of bacteria to survive in close proximity to certain moulds looked trivial at first, but few informed people would label the discovery of that initial fact *trivial* today."

So much, then, for a few illustrations, rather than a summary, of the type of work that is being done and that needs to be done in the social sciences.[6] Is there enough of it being done? Clearly not, or we would not need to flounder as we are in national and international affairs, pursuing diametrically opposite courses within the same decade. Can the social sciences ever hope to catch up with the other sciences, the increasingly rapid advance of which constantly creates new social problems? Certainly we can, if we devote ourselves to the business with something like the seriousness, the money, and the equipment that we have devoted to physical re-

[6] See also Chapter VI, Section IV.

search. Consider how the physical scientists are today given vast resources to concentrate on the invention of a new submarine detector or a new bomb, not to mention the peacetime occupations of these scientists with penicillin and sulpha drugs. Obviously, I am not criticizing this action. On the contrary, it is the way to proceed if you want results. Is there anything like that going on regarding the world organization and its numerous subsidiary problems, all of them important to peace and prosperity?

Comparatively speaking, there is almost nothing that could be called fundamental research into the basic nature of human relations. To be sure, there are endless petty projects, surveys, conferences, oratory, and arguments by representatives of pressure groups, as if argument ever settled any scientific questions. Of basic social research there is very little.[7] Why isn't there more? As we pointed out in the first chapter, it is not yet realized that scientific knowledge is relevant to successful world organization. We still think that common sense, good will, eloquent leaders, and pious hopes are sufficient when it comes to management of social relations.

III

The Cost of a Transition to Science

This brings us to our second question. What price must we probably pay for a social science of a comprehensiveness and reliability comparable to some of the better developed physical sciences? The costs are undoubtedly considerable, and it remains to be seen to what extent men are willing to

[7] For an excellent brief summary of the recent trend toward more adequate moral and financial support of the social sciences by the federal government, see H. A. Alpert, "The Government's Growing Recognition of Social Science," *Annals of the American Academy of Political and Social Science,* January, 1960, pp. 59–67.

pay them. What are some of the principal items both as regards material and psychological costs?

The mention of costs suggests that I am about to digress into the subject of research finance. The advancement of science undoubtedly does involve costs of this type. I shall not go into them here, because I am at present more concerned with other types of costs which have nothing to do with money or with budgets. Let me therefore dismiss the question of monetary costs with a brief estimate by Huxley: "Before humanity can obtain," he says, "on the collective level that degree of foresight, control and flexibility which on the biological level is at the disposal of human individuals, it must multiply at least ten-fold, perhaps fifty-fold, the proportion of individuals and organizatioins devoted to obtaining information, to planning, to correlation and the flexible control of execution."[8] This may seem staggering to educators who are wondering how to maintain merely their present activities. But how does the entire expenditure for scientific research compare with what we have spent and are spending on war? Perhaps it will occur to some future generation to try a reallocation of public funds. If so, adequate research and training in social science can be readily financed.

But are we or is some future generation likely to change so radically our notions of what is worth spending money for? This brings us face to face with those costs of science which perhaps come higher, and touch us more deeply, than any of its financial costs.

First of all, the advancement of the social sciences would probably deprive us in a large measure of the luxury of indignation in which we now indulge ourselves as regards social events. The "cold war" is a case in point. Such indignation ministers to deep-seated, jungle-fed sentiments of justice, virtue, and a general feeling of the fitness of things,

[8] J. Huxley, *op. cit.*

as compared with what a scientific diagnosis of the situation evokes. In short, one of the principal costs of the advancement of the social sciences would be the abandonment of the personalistic and moralistic interpretation of social events, just as we had to abandon this type of explanation of physical phenomena when we went over to the scientific orientation.

Closely related and indeed inseparably connected with the necessary abandonment, in science, of personalistic and moralistic types of explanation is the necessity of abandoning or redefining a large vocabulary to which we are deeply and emotionally attached. Concepts like freedom, democracy, liberty, independence, free speech, self-determination, and a multitude of others have never been realistically analyzed by most people as to their actual content under changing conditions. Any such analyses, furthermore, are sure to seem like an attack upon these cherished symbols and the romantic state of affairs for which they stand. As every social scientist knows, these are subjects that had better be handled with care.

Or consider the opposite kind of words, such as "aggression" or "world conquest." In 1940 we were greatly agitated over the aggressions and alleged clear intentions of world conquest on the part of Germany. By 1948 we were convinced that Russia was the master aggressor and world conqueror. A little later the Russians and their friends found that when it came to aggression and world conquest the U. S. A. was the obvious villain. The United Nations at one time undertook to define aggression but abandoned the attempt. It was finally voted that aggression was "a concept per se [sic], which is not susceptible of definition." *(Bulletin of the Atomic Scientists,* Vol. 9, April, 1953, p. 69.)

Social sciences worthy of the name will have to examine realistically all the pious shibboleths which are not only frequently the last refuge of scoundrels and bigots, but also serve as shelters behind which we today seek to hide the facts we are reluctant to face. The question is, how much pain in the way of disillusionment about fairy tales, disturbed habits of thought, and disrupted traditional ways of behavior will the patient be willing to put up with in order to be cured of his disease? He will probably have to become a lot sicker than he is before he will consent to take the medicine which alone can save him.

Finally, the advancement of the social sciences will cost the abandonment not only of *individual concepts* carried with us from prescientific times, it will require us also to abandon deeply cherished *ideologies,* resembling in form, if not in content, their theological predecessors. The notion of some final solution, preferably in our own generation, of the major social problems that agitate us is a mirage which even scientists have great difficulty in abandoning. Many of them still confuse the social sciences with various cults, religions, and political dogmas, from Marxism to astrology. Scientists must recognize that democracy, for all its virtues, is only one of the possible types of organization under which men have lived and achieved civilization.

It is a disservice to democracy as well as to science to make preposterous claims that science can prosper only under some particular form of government, that only under our particular form of political organization do minorities have rights, etc. The favorite cliché is that "science can flourish only in freedom." It is a beautiful phrase, but unfortunately it flagrantly begs the question. The question is, under what conditions will the kind of freedom science needs be provided? The historical fact is that science has gone forward under a great variety of forms of government, and conversely, at other times, has been suppressed and frustrated by each of the same types, including democracy. The first truly popular democratic government in Europe, namely, the

French Revolution, declared itself to have "no use for scientists" and proceeded to behead Lavoisier, the father of modern chemistry. Only a few decades ago, several states, under the leadership of American "statesmen," passed laws against the teaching of evolution. American citizens of Japanese ancestry have even more recently discovered precisely what the Bill of Rights amounts to in a pinch, especially under an administration and a Supreme Court eloquent in their verbalizations about the rights of minorities. In short, the great democratic "gains" in this department appear to consist, in the opinion of qualified legal analysts,[9] of creating a legal status for minorities in the United States somewhat comparable to that which they had in Nazi Germany.

In short, attacks both on science and on "freedom" *do occur* also in democracies. I would condemn them *wherever* they occur. The attempt to make science the tail of *any* political kite whatsoever must be vigorously opposed by all scientists as well as by all others who believe in uncorrupted science. Political systems have changed, and they will change. Science has survived them all as an instrument which man may use under any organizatioin for whatever ends he seeks.

The mere fact that I, personally, happen to like the democratic way of life with all its absurdities, that I would find some current alternatives quite intolerable, and that I may even find it worth-while to go to any length in defense of democracy of the type to which I am accustomed are matters of little or no importance as touching the scientific question at issue. My attachment to democracy may be, in fact, of *scientific* significance chiefly as indicating my unfitness to live in a changing world. To accept this simple notion is perhaps a cost of social science that few are prepared to pay.

[9] See E. V. Rostow, "Our Worst Wartime Mistake," *Harper's Magazine*, September, 1945.

IV

The Task of Social Scientists

We have reviewed above some of the costs that are involved in the transition to a scientific view of human relations. These costs affect scientists as well as other people. In addition, scientists have one special concern. What does the future promise for social scientists in the way of freedom to perform the tasks which I have outlined as their proper business?

Social scientists need not expect to escape the troubles which other scientists have encountered throughout history. Chemists and physicists from time to time have suffered persecutions because of the conflict of their findings with more generally accepted views. They have continued to hew to the line, however, until today they enjoy a certain immunity and freedom of investigation which social scientists do not share. Why do physical scientists enjoy this relative security in the face of changing political regimes, and how may social scientists attain a corresponding immunity?

The answer is popularly assumed to lie in the peculiar subject-matter with which social scientists deal. I doubt if this is the principal reason. I think a far more fundamental reason for the relative precariousness of the social sciences lies in their comparative incompetence.

Social scientists, unfortunately, have failed as yet to convince any considerable number of persons that they are engaged in a pursuit of knowledge of a kind which is demonstrably true, regardless of the private preferences, hopes, and likes of the scientist himself. *All* sciences have gone through this stage. Physical scientists are, as a class, less likely to be disturbed than social scientists when a political upheaval comes along, because the work of the former is recognized as of equal consequence under any regime. Social science should strive for a similar position. Individual physicists may suffer persecution, but their successors carry on their work in much the same way. If social scientists possessed an equally demonstrably relevant body of knowledge and technique of finding answers to questions, that knowledge would be equally above the reach of political upheaval. The services of *real* social scientists would be as indispensable to Fascists as to Communists and Democrats, just as are the services of physicists and physicians. The findings of physical scientists at times also have been ignored by political regimes, but when that *has* occurred, it has been the *regime* and not the *science* that yielded in the end.

The trivial effect of political interference upon the well-developed sciences should be noted. I recognize, of course, the frequently unfortunate effect of these movements upon individual careers and individual projects. But, if we plot the course of scientific advance during the past two hundred years, the impressive fact is how little its main course has been deflected by all the petty movements of so-called "social action," including the major political revolutions. The demonstrable superiority of science as a method of achieving *whatever* men want has caused even its persecutors to return to it, after only very temporary and superficial attacks, chiefly against individual scientists. The recent history of Germany and Russia are cases in point.

I have emphasized that physical scientists are indispensable to any political regime. Social scientists might well work toward a corresponding status. Already some of them have achieved it to a degree. Qualified social statisticians have not been and will not be disturbed greatly in their function by any political party as long as they confine themselves to their specialty. Their skill consists in the ability to draw relatively valid, unbiased, and demonstrable conclusions from observed data of social behavior. *That* technique is the same, regardless of social objectives. No regime can get along without this technology. It is the possession and exercise of such skills alone that justifies the claim of academic immunity. To claim it for those who insist on taking for granted that which needs to be demonstrated can only result in the repudiation for everybody of the whole principle of academic freedom. For the same reason, we had better not become so devoted to blatant crusades for academic freedom that we forget to bolster the only foundation upon which academic freedom can ever be maintained in the long run, namely, the demonstrated capacity of its possessors to make valid and impersonal analyses and predictions of social events.

The temptation is admittedly considerable to bolster one's favorite "movement," by posing as a disinterested appraiser of the truth while actually engaging in special pleading. It is also tempting in this way to seek the right of sanctuary in the form of academic freedom to escape the ordinary consequences of pressure group activity as visited on less clever and less privileged people. Special pleading must be recognized for what it is whether it serves the Congress of Industrial Organizations or the National Association of Manufacturers. I have no objection to universities maintaining forums for special pleading nor do I object to scientists taking part in such discussion as long as no attempt is made to pass the whole thing off as "science." Too frequently scientists forget this distinction and put forward absurd scientific claims for what they personally happen to prefer.

The form of social organization which will yield to men the satisfaction they desire obviously depends upon a great number and variety of factors, including traditions, resources, technology, scientific development, and education. Scientists would do better to make it perfectly clear that their personal preferences in these matters are merely their own current preferences and not scientific conclusions valid for all times and places or conditions of people.

My conclusion, then, is that the best hope for the social sciences lies in following broadly in the paths of the other sciences. I have not tried to minimize the difficulties that beset these paths. I have merely argued that they are not insurmountable, and that in any case we really have no choice but to pursue this one hope. For we are already so heavily committed to the thoughtways and the material results of science in so large a part of our lives that we are likely to go farther in the same direction. In short, the trends that have been strikingly evident in the social sciences in recent decades will, I believe, continue at an ever more rapid rate. Social scientists will talk less and say more. They will rely ever more heavily on a more economical type of discourse, namely, the statistical and mathematical. Much of what now passes for social science will be properly relegated to other equally honorable departments, such as journalism, drama, or general literature. As such, this material will have its uses as propaganda, news, art, and a legitimate outlet for the emotions of men. Indeed, nothing I have said regarding the possibilities of scientific study of human affairs should be interpreted as in any way contemplating an abandonment or a restriction upon the artistic, religious, literary, or recreational arts which also minister to the cravings of men. I have on the contrary rather advocated that the social sciences should not handicap themselves by aggrandizing to themselves roles which they cannot fulfill.

Social scientists, as scientists, had better confine themselves to three tasks: First and foremost, they should devote themselves to developing reliable knowledge of what alternatives of action exist under given conditions and the probable consequences of each. Secondly, social scientists should, as a legitimate part of their technology as well as for its practical uses, be able to gauge reliably what the masses of men want under given circumstances. Finally, they should, in the applied aspects of their science, develop the administrative or engineering techniques of satisfying most efficiently and economically these wants, regardless of what they may be at any given time, regardless of how they may change from time to time, and regardless of the scientists' own preferences.

V

Is Science Authoritarian?

We have now reviewed, first, some of the things the social sciences have done, some of the things they are trying to do, and some of the things they need to do. Secondly, we have considered some of the costs, both in money and in altered habits of thinking and feeling, which the transition to a scientifically ordered society involves.

What further consequences might we expect in a transition to a scientific view of human relations? The question has been raised as to whether the above proposal contemplates that every citizen should become a social scientist. On the contrary, the proposal is rather to relieve the ordinary citizen of many impossible duties with which he is now saddled and which, because they are impossible, he refuses to perform. The development of the physical sciences did not require every person to become skilled in every science. We do try to teach every person that there *is* such a thing as science, and that it is better for him to consult its duly licensed practitioners rather than to doctor himself. To be sure, this is not an infallible system, but it works pretty well. I contemplate a similar relation with respect to social science.

Will the citizen, then, be expected to lean more heavily than at present on the conclusions of social scientists? Well, he can do as he pleases, but I notice that he pleases to accept rather blindly the authority of physical scientists.

A similar attitude toward the conclusions of social scientists is suspected of being authoritarian, as indeed it probably is. A lot of nonsense has been spoken and written about authority in recent years. We need to recognize that it is not authority as such that we need fear but incompetent and unwisely constituted authority. When we undertake to insist on the same criteria of authority in the social as in the physical sciences, no one will worry about the delegation of that authority, any more than he worries about the physician's authority. All persons who presume to speak with authority will be expected to submit credentials of training and character of the type that physicians and other professionals now submit, and *to the state,* at that. This will hold for all would-be authorities whatsoever, whether they purport to speak for God or for nature. The state may in turn be required to *delegate* the function of formulating and administering these requirements back into the hands of members of the profession concerned. This does not alter the importance of retaining the ultimate authority in the hands of the community's accredited governmental agency.

The notion that things are quite democratic, provided you hold enough elections, conventions, and meetings, seems to be taken for granted by many people.[10] Some of the most ardent defenders of democratic processes are finally waking up to the fact that in clinging to certain traditional democratic practices we may be retaining the shadow and sacrificing the substance of democracy. For example, that great and consistent champion of the rights of organized labor, The American Civil Liberties Union, in its *Bulletin on Democracy in Trade Unions,* has this to say:

"It is clear that no reforms in union constitutions, rules

[10] The get-out-the-vote agitation is based on this supposition. For a contrary view, see R. E. Coulson "Let's Not Get Out the Vote," *Harper's Magazine,* November, 1955.

and regulations will in themselves achieve trade union democracy and insure against undemocratic practices. No rules or regulations will stop packing meetings with the followers of a group determined to achieve power; they will not stop strong-arm methods of gangsterism; they will not prevent a minority from achieving control by outsitting the majority at interminable meetings, nor prevent a determined minority from getting together in a caucus and voting en bloc after dividing its opponents."

Obviously, this may be equally true of employers' organizations or any other type. Clearly, we need new social inventions, a new technology for making the public will known and operative, without involving the abuses mentioned. This will doubtless be opposed by those who profit from the present malfunctioning of democratic methods. Unfortunately, the most eloquent verbal defenders of democracy are frequently its worst enemies.

Under these conditions, the common man will correctly look for other means of guarding his interests. He assures himself today as far as possible against malpractice on the part of engineers, electricians, doctors, lawyers, and teachers by requiring them to qualify according to state-regulated criteria. He thereupon gives his authorities and technicians a free hand and holds them responsible for results. Most of the multifarious duties of the private citizen today will, I predict, go this way in the not too distant future. For some time past, the "ward heeler" has functioned as a sort of informal, unofficial social worker to relieve the citizen of some of his obligations as a citizen. This can happen anywhere on a national scale when the pressure becomes sufficient. If scientists do not assume their proper responsibilities, charlatans are likely to take over.

The trend mentioned above is merely a transition toward the responsible performance by scientists, under the auspices of the public authority, whether local, national, or international, of certain functions that were hitherto imposed upon each citizen. Now, one may have an open mind as to the proper or desirable function of the state, because these questions depend upon how the state is constituted and upon the level of scientific development that a society has achieved. But there can be no doubt at all that the authority of a *properly constituted* state is preferable to what seems to be the alternative, namely, private and self-constituted legislatures, police, and courts, as they occur today in many countries, seeking to impose their private wishes upon the larger public.

This state of affairs is quite natural and perhaps fortunate in the sense that some leadership or some solution is better than none. When people are in trouble, they will look for a saviour. Now there are certain temperaments in all countries which enjoy action on the basis of guess, magic, astrology, or intuition. These are likely to come into power especially in periods of crisis. They are likely to surround themselves with seers, poets, playwrights, and others alleged to possess these power of "seeing." The idea of foresight is a sound one. The only reform needed is a substitution of scientists for these soothsayers and soothseers, always remembering the important difference between *"foresight,"* in the sense of reliably predicting costs and consequences, as contrasted with *dictating* which of several possible choices should be followed.

Fortunately, distinguished physical scientists are also beginning to take that view of the matter. This is the more important in view of the fact that the primitive sociological views of some of the prominent physical scientists have hitherto been an obstacle to the development of a hospitable attitude toward social science in the very places from which the most influential and valuable technical support should be forthcoming. Encouraging, therefore, is the following statement from Dr. Frank B. Jewett, President of the National Academy of Sciences, in collaboration with Dr. Robert W. King. After reviewing the need in statecraft for something corresponding to the research laboratory in industry, these scientists conclude:

"It seems likely that we are well launched upon an era during which all the existing advisory aids to the government, as well as others still to be created, will have to function with increasing vigor. Such an arrangement will not savor of bureaucracy. The sovereign people will still remain sovereign. But belated and constructive recognition will have been given to the fact, now abundantly clear, that the day is gone, and probably forever, when a successful state can base its policies upon the clamor of pressure groups or upon the uninformed beliefs of the majority, even though measured numerically by tens of millions."[11]

Scientists in general are finally awakening to the fact that unless the social sciences are developed our net reward for the development of the other sciences may be destruction. When that is fully realized, we may expect the same kind of support for the social sciences as the physical sciences have received and from the same sources and for the same reasons. Corporations have for some time been adding sociologists to their staffs, because what they want from business and from life is to be secured only through an adjustment of human relations in the factory as well as in the community, the nation, and between nations. Social scientists do not compete with physical scientists, but complement them. The further development of mechanical technology is of doubtful interest if the net result is to lay waste the countryside with bombs. When that fact is realized, institutes of social research will rise in every university and in every

[11] "Engineering Progress and the Social Order," University of Pennsylvania Bicentennial Conference, Philadelphia, September 19, 1940.

large industrial establishment. Social scientists will not be expected to perform miracles. They will merely be expected to bring to the study of man's social behavior the same objectivity, relevant skills, and devotion to scientific truth which have in the past given us some understanding and control of the physical world.

CHAPTER IV

Education in a Scientific Age

I

The "Literary" Approach

WE HAVE been dealing in these chapters with a matter which transcends in importance the ups and downs of depressions, contemporary politics, and war. That matter is man's constant struggle through the ages to arrive at a method of approach in thought and action which will be relatively valid no matter what ends man elects to pursue and no matter through what form of political organization it is directed. Consider the forms of government that have been dominant from time to time during the last four hundred years. They have varied from extreme autocratic monarchies and dictatorships to the extreme democratic ideas of the French Revolution. But running consistently through these revolutions, movements, and ideologies, many of them aimed at quick and easy salvation, has been the steady advancement of science. To it, rather than the much more publicized social and political movements, has been due the improvements of man's lot.

Today, however, men cry out that while science has indubitably brought us toward some of our goals, it has not brought us that peace and orderliness in human relations which we desire above all. The main point of these chapters has been that this result can come only when we develop and apply the sciences dealing with human relationships. This argument takes us to the subject of education and research. That is, if the scientific method, and especially its application to human relations, is as important as we have contended, then our educational efforts must be judged largely by the degree to which they inculcate a familiarity with this method, and the reliable generalizations it has yielded thus far.

Judged by this standard, research and education in the social sciences have as yet hardly begun. It is true that a great deal of time and a large part of the curriculum is devoted with varying degrees of directness to the subject of human relations. In fact, nearly all teachers, no matter what subject they teach, make detours into sociology, perhaps chiefly to air their own prejudices. Outside of the schools, the prominence of this subject is even greater. Never in history have so many and so eloquent words been printed and circulated on social questions. Innumerable forums, panels, lectures, round tables, and radio commentators are continuously churning the atmosphere with verbalizations about sociological subjects. Almost none of this discussion is in the framework of science. Almost all of it is in legalistic-moralistic terms resting directly upon other than scientific assumptions regarding human behavior.

Now, obviously, when I speak disparagingly of the legalistic-moralistic approach, I am not reflecting upon either law or morality as such. Laws and morals are as necessary and as desirable in a scientific as in any other kind of world. All I am saying is that we must not retain superstitious notions about the nature of these rules or the manner in which they originate in human society. Neither must we mistake verbalizations about ideal laws and morals for people's actual behavior as it is, nor overlook the conditions which produce both law and morality as we find them.

I have illustrated in preceding chapters how this failure to distinguish fiction from fact gets us into all kinds of insoluble situations. One more illustration may be permissible at this time, from the field of international relations, because of the current interest in that subject. The late Professor

Nicholas J. Spykman of Yale in his book, *America's Strategy in World Politics,* summarized the point as follows:

"The heritage of seventeenth-century Puritanism," he says, "is responsible for one of the characteristic features of our approach to international relations. Because of its concern with ethical values, it has conditioned the nation to a predominantly moral orientation. It makes our people feel called upon to express moral judgments about the foreign policy of others and demand that our president shall transform the White House into an international pulpit from which mankind can be scolded for the evil of its ways. The heritage of eighteenth-century rationalism has contributed another characteristic feature, a legalistic approach, and a faith in the compelling power of the reason of the law. This almost instinctive preference for a moral and legal outlook on international affairs tends to obscure for the American people the underlying realities of power politics."

No nation has such a good time moralizing as we do. The thoughtway is a simple, not to say a simple-minded one. Men and nations *shouldn't* behave in certain ways, therefore, it is believed, they *won't* behave that way, especially if we exhort them enough. Although the whole course of history indicates the fallacy of this reasoning, we cling to it. Our actions are correspondingly naive. We sign treaties with all nations agreeing to give up war as an instrument of national policy, and then relax as if war had been made unlikely. The premises and the reasoning are very much like those underlying *magical* rain-making. That is, we *want* it to rain, therefore it *should* rain, therefore it will rain. We have discovered the invalidity of this reasoning in the case of rain, and our schools for the most part no longer teach magical methods of influencing physical events.

The schools, however, still teach similar doctrines regarding human relations. For man, it is held, unlike the rest of nature, does what he does by an act of will, or at least acts in obedience to good or bad forces *outside* this world. It is not necessary here to get into an argument over free will. *All that the scientist wishes to do is to establish the laws describing the conditions under which man wills the way he does.* Moralistically, it is pointed out, for example, that Germany didn't *have to* invade other countries; that she *could have* gone in for butter instead of guns; that Germany *had* or *could have had* a higher standard of living than some of her neighbors; etc. The scientist finds all this somewhat irrelevant. He is interested rather in observing the various conditions under which nations do engage in war. He finds that our own expansion, and that of every one of the so-called peaceful and freedom-loving nations have not been cases of low-standard people preying on higher-standard people but quite the contrary. This observed fact and, of course, many others are to the scientist the relevant considerations with which we have to deal.

In short, we still draw in our education a firm line dividing man and his affairs from the rest of nature. To phenomena on one side of this line, we apply such words as "materialistic," "physical," "natural," "sensory," etc. To phenomena on the other side, we apply words such as "non-material," "social," "mental," "voluntary," "willful," "spiritual," and the whole mentalistic vocabulary. This split, to which I called attention in the first chapter, constitutes the "basic problem of present culture and associated living." It is this problem that cannot be solved apart from a unified method of attack and procedure. That method must be the method of science. This diagnosis, unfortunately, has not yet made any considerable impact upon our school system.

I suggested in the preceding chapter that much of the resistance to an adequate recognition of the role of science in human relations is due to a misunderstanding of the relation of social science to the other departments of life. When the nature of social science is better understood, there will be few who will want to throw mankind forever back into the brutish abyss of the animistic and supernaturalistic conception of himself and the universe. Already some elements of the opposition have abandoned the traditional animistic and even the theological view and have adopted a kind of middle ground which rests its case upon a certain body of lore called "the classics." It is significant to note that only in regard to human relations are "the classics" invoked. Strangely, no one suggests that these same "classics" should be our guide in engineering, medicine, or agriculture. In these fields the most confirmed classicist agrees that we had better use the latest accredited findings of the research laboratories.

It is only a question of time until we shall take the same view regarding the authority of the classics in human relations.

II

The Content of Education

What, then, should be the content of general education today? Perhaps most people would agree that the purpose of general education is to develop an ability to cope with the world to the maximum satisfaction of the student himself and of his community. In order to avoid needless controversy, I shall consider here education only in the lower schools, the high schools, and perhaps the first two years of college. Higher, specialized education for the few will be largely omitted from consideration in this discussion.

It still remains a question what part of the whole educational task should be undertaken by the schools, and what part should be left to other institutions, such as the home, industry, and the various recreational agencies. Here we may perhaps agree that teaching vocational subjects is not the *primary* function of the liberal arts college, although these subjects are admittedly most vital in a person's adjustment in the social order, and certainly should be taught somehow, somewhere. What is to be taught in school and what is to be taught through other community agencies should be decided on the basis of what yields the most satisfactory results.

For the content of general education we may use as a guide the relative consensus with which certain subject matter is most generally included in the curriculum of the lower schools. Curricula vary greatly in their detailed offerings, but there is very wide consensus regarding the teaching of such elements as reading, writing, arithmetic, and the rudiments of physical and social science in the elementary schools. From this I conclude that we are already convinced of the importance of this general subject matter and are therefore warranted in maintaining a curriculum requiring proficiency in the arts of verbal communication (including mathematics) and in science (both physical and social). Further, I think there would be wide agreement that an education should include some familiarity with those subjects and behaviors to which men have always looked for refreshment, release, creativeness, and pleasure, namely, the arts. A general educational curriculum in our culture, then, should have at least two broad divisions, science, including the technology of denotative communication, and art.

History, philosophy, and literature include perhaps the bulk of the subject matter usually referred to as humanistic. It would greatly clarify and resolve certain controversies if this subject matter were reclassified into the categories of science and art. Traditional philosophy, for example, consists primarily of the history of ideas, including metaphysics, and is therefore classifiable as history. Logic belongs with mathematics, grammar, semantics, etc., some of which are already well recognized as essential scientific tools, and as such are logically a part of science, if not sciences in their own right, just as mathematics is everywhere recognized as a tool of science. A new science of semiosis, dealing with the nature of symbolic behavior, is in process of emerging, and will be as important, I think, to the social sciences as mathematics has been in the evolution of the physical sciences. All of these tool subjects represent no problem of classification in the curriculum of a general education, because they should be learned like reading—i.e., as required tools in mastering either science or art.

History is in a peculiar position in that it might include everything or nothing. Its principal subject matter is already largely included in, and is in any event logically part of, cultural anthropology, which is indistinguishable from sociology. Both sociology and cultural anthropology perform the "synthesizing" function which is sometimes alleged to be the unique contribution of history. Furthermore, history is taught by everyone, since all scientific and artistic data are necessarily historical. The history of music, of art, of physics, of culture, of ideas, of politics, of institutions, are all taught by the people in these respective fields. There is no residue of "history," as such. It is always the history *of something.*

History will doubtless continue as a separate discipline for a long time, as a valuable adjunct to, and finally as a necessary part of, the respective social sciences. In the long-overdue separation of these sciences from journalism, literature, propagandist oratory, and drama, substantial parts of history will of course be classified as literature. Objective historical fact, being the raw material of all science, will continue as a major part of the curriculum. The collection and interpretation of this material is the concern of all the sciences. The present discussion, therefore, casts no aspersion upon historical data. I am discussing here only the logic of the present organizatioin of history as a special subject in the curriculum.

Those types of historical writing in which mythology and

literary style are the dominant considerations clearly belong with literature, which, in turn, will increasingly accept its role as an art. This will be to its own great advantage. Music and other arts have prospered much and are on a securer foundation for abandoning all pretense at curing smallpox, producing fertility of the soil, or depicting current events. Literature likewise will find it advisable, I think, to pose less and less as an authoritative spokesman on psychology, sociology, economics, and foreign policy, and concern itself chiefly with the creation of, and conformity to, *artistic* standards. This appears to be what President Clark Kerr had in mind when he said:

"If the humanities are dying, I am not convinced that it is a case of murder or that the administrators are the most likely suspects. It seems to me that in a good many departments, in a good many universities, it is a case, rather, of attempted and as yet not entirely successful suicide. Now, it seems to me that one of the troubles with the humanists is that they are trying very hard to be something that they cannot be. They are trying to be scientific. And they turn with a vengeance to history and develop all sorts of unimportant facts about unimportant poets, or they develop pseudo-scientific criticism, and I have wondered a bit why this was so frequently true, and wondered whether this wasn't an effort to get the prestige that goes along with science and also the protection that goes along with developing an independent discipline, which nobody else understands, and thus nobody else can properly criticize. The humanities no longer teach the classics as they once did. They are no longer willing to go through the routine problems of improving style; even the language professors are not willing to teach language, but reserve their best efforts for minute points of textual criticism, and the like. So, it seems to me, the humanities shouldn't be what they can't be. They don't have to be exact, they just have to be interesting and colorful and delightful. They can be speculative, concerned with values, and thus they can really be useful and of service to everybody ... the sin of the humanities is that they are now trying to become women of virtue, correct and proper and highly isolated, and they would make a greater contribution if they would return to their earlier and more alluring ways."[1]

This obviously does not mean that art does not or cannot portray human relations. It merely means that the objective validity of that portrayal is a question for scientific determination. If our notions of human nature and relations are themselves, as at present, largely derived from folklore and literature, then whatever art confirms these notions seems to "ring true" and thus appears to be validated. *Scientific validation* of the alleged contributions of literature to psychology and sociology waits upon scientific test. This is precisely what is needed because social relations are today managed on the basis of what poets, playwrights, journalists, preachers, and radio commentators assume, on the basis of folklore, literature, and highly limited personal experience, to be principles of human nature and human relations. In that testing, a certain degree of validity will undoubtedly be found in Homer, Shakespeare, and others—perhaps about the same validity as has been found in the geographical, physical and biological contributions of these authors. We shall return to this subject in the next chapter. (See pp. 92–96.)

A reasonable content for general education today, then, seems to me to be as follows: First, a command of the principal linguistic tools essential to the pursuit of either science or art. Second, a familiarity with the scientific method and with its principal applications to both physical and social problems. And third, appreciation and practice of the arts, including literature. Furthermore, these three fields should

[1] Clark Kerr, *The Educational Record,* January, 1955, pp. 69–70.

be so integrated toward a common purpose that the question of their relative importance would not even arise. One does not ask which is the most important leg of a tripod.

III

Principal Defects

In view of the above requirements of a general education in the modern world, what are the principal defects in our educational program as it stands?

1. The first defect of our present educational program is fortunately one on which all seem to agree, including even Mr. Hutchins and Mr. Dewey. I, too, find myself in full agreement with Mr. Hutchins' *diagnosis* of the situation and differ with him only about the *remedy*. I agree with him that modern education suffers from the lack of a unifying discipline, such as theology provided in the Middle Ages. I contend, however, that this unifying discipline today cannot be, as some feel, an amorphous mass of one hundred or any other stated number of books, called "the classics," dealing primarily with the ancient and medieval gropings of mankind to find its way about. The unifying discipline of modern education lies in modern science, which is simply the distilled essence of the classics of all time. Science is the positive residue of the sifting and winnowing of all the ages, and its methods are the demonstrably effective instruments of reaching valid conclusions. In this respect, and in the fields where it has been tried, science has superseded theology and philosophy as the acknowledged authority. Our reluctance to accept a similar transition of ways of thinking regarding social phenomena is at the root of the confusions, the conflicts, and the disunity which we find even among the highly educated.

2. If we agree that our failure to recognize the dominant

importance of the scientific method is our first great oversight, then the second principal defect of our educational program is clearly our failure to develop, even in our college students, an adequate knowledge of the elements of that method. I called attention in an earlier chapter to the absurdity of calling the present a scientific age, in view of the way in which we think, talk, and act about human relations. How does this happen to be so in view of the tremendous vogue of science in some departments of our life, the excellent scientific equipment in many of our schools, and the prominence of the subject in public discussion?

There are three principal explanations: First of all, there is the deep-seated notion, mentioned in the first chapter, that science has to do exclusively with the physical world and that its methods are not applicable to human affairs. Why is it that we find ourselves with thousands of research laboratories devoted to the application of the scientific method to physical problems, and almost none, comparatively speaking, devoted to research in human relations? Is it because human relations are functioning so smoothly that they need no attention? A glance at industry, our minority problems, and the international situation should be sufficient answer. Is it because, while physical research pays a tangible dividend in profits, improvement in human relations does not? Ask any employer what labor troubles, strikes, lockouts, slowdowns, and loafing is costing him, and in the end, costing all of us. Then why don't we risk a few million dollars in scientific research on human relations the way we risk hundreds of millions on industrial research, because in the end we know it will pay? There can be only one answer: We do not yet believe that human relations is a subject to which the scientific method is applicable.

To be sure, there are a few professors stealing some time from their teaching to do some research. To be sure, industries have their personnel departments, efficiency engineers, social workers, and trouble shooters. Many of them are clever and useful people. But they are, in the nature of the case, about as helpless as the physician was before the sciences of chemistry, physiology, and bacteriology developed. To be qualified to pull a tooth or remove an appendix, we require people to study systematically for seven or eight years beyond high school. To keep nations from flying at each other's throats, any political hack will do. Human relations will improve when we undertake serious scientific study of how to improve them. In the meantime, we continue to rely on incantations, denunciations, exhortations, and exorcism exactly as our prescientific forefathers did regarding their physical maladjustments.

A second reason for the failure of our schools to turn out people more familiar with the scientific approach is that even the physical sciences still receive a minor proportion of attention in the liberal arts college, not to mention the grades and high school. It is true that there is a vast amount of scientific education going on in industry and trade outside of the regular public schools and colleges. It is also true that the current excitement about sputniks, and space travel has brought on a voluminous discussion which has already forced some improvement in the recently existing state of affairs. But the hue and cry that has been raised in some quarters about science running away with our schools is clearly not warranted. The fact is that instruction in science of all kinds in our schools is still lagging badly behind the needs of our present culture.

The third cause of the scientific illiteracy of a generation enmeshed in the material toils and consequences of science is the type of scientific instruction that is offered. Such is the preoccupation in the teaching of science with specific subject matter and with specific technologies, that many students, although they major in some particular science, have no conception of the scientific method as a generally valid approach to the problems of this world. Science is generally thought of as a type of *subject matter* rather than as a *method* of study. To some, science means colored liquid in a glass tube; to others, the paraphernalia of the physics laboratory; to still others, it signifies a terminology liberally interspersed with mathematical formulas. The idea of science as a particular method of study, a definite set of rules of procedure and of logic, applicable to *any* subject matter, has been neglected in our preoccupation with the startling findings and achievements of the better developed sciences. Hence it is not uncommon for a science "major" to discover for the first time the meaning of the scientific method in a course of philosophy, history, or sociology.

Actually, scientific method, as such, probably should be taught in courses in philosophy or sociology because science, as a method, is a form of human behavior. It consist of "(a) asking clear, answerable questions in order to direct one's (b) observations, which are made in a calm and unprejudiced manner, and which are then (c) reported as accurately as possible and in such a way as to answer the questions that were asked to begin with, after which (d) any pertinent beliefs or assumptions that were held before the observations were made are revised in light of the observations made and answers obtained."[2] Scientific *method,* therefore, is perhaps best taught in subjects dealing with the nature and development of human thought, including logic, as the system of rules governing specified kinds of thinking. In view of this fact, it is perhaps unfair to expect chemists, physicists, and biologists to teach scientific method. Many of them are not especially qualified for this work. They already have a very heavy assignment in teaching their own subject matter and, at most, in showing what scientific method means in their own specialized fields. If so, obvi-

[2] W. Johnson, *People in Quandaries* (New York: Harper & Bros., 1946), p. 49.

ously, we should stop blaming them for failing to do something which they are not trying to do and probably haven't time to do anyway, if they are to do the other things we expect of them.

In short, if we wish to remedy the present failure of our schools to acquaint students with even the elements of scientific method, we must establish and require, from the grades through high school and college, courses definitely calculated to acquaint everyone with the broader meaning and methods of science. Especially must we extend the application of that method to the realm of human social behavior.

The question at once arises as to who would teach these new courses, in what department they would fall, etc. That is a question of administrative detail with which it is not necessary for my purpose to deal, except to say that, like other courses, they should be given by those best qualified to teach them. I am aware, of course, that some philosophy departments have long been teaching courses in the history and logic of scientific method, that history and philosophy departments have long been teaching the history of ideas, including science, and that elsewhere sociologists have been compelled to teach these subjects because no one else was doing so. I am not at all interested at present in departmental jurisdiction, but I am mightily interested in getting this subject taught somehow, somewhere, to all people who are to be regarded as literate in the present world. That it is not being adequately taught today is beyond question.

The gullibility of vast numbers of otherwise apparently intelligent people to the propaganda, quackery, and improbabilities of the day is perhaps the best index of the degree to which scientific habits of thought, as contrasted with a lot of facts *about* science, have penetrated. So superficial is the habit of rational and scientific thinking about social events that in situations at all critical (e.g., war), even the scientifically trained promptly abandon this comparatively recent acquisition in favor of a more primitive and emotional personalistic method which the conditionings of childhood, and generations of prescientific thinking have ingrained in us in a way that it will take centuries for science to overcome. Yet this is the principal and most urgent task which faces our educational institutions. To the extent that it is achieved we shall be masters of our social as well as our physical world.

3. The third great defect of our educational program at present is its failure to equip students with an adequate understanding of the nature and uses of language. Just as it is impossible to teach physics without some understanding of mathematics, so it is impossible to teach any social science without an understanding of the particular kind of language usage and logic which all science absolutely requires.[3] Indeed, the first lessons in scientific methods, in their broader meaning, as contrasted with mere laboratory techniques and scientific information, must be concerned with the nature and uses of language and logic. Adequate instruction in the fields of linguistics and semantics would contribute much to the resolution of the thorougly obsolete cleavage between the "mental" and the "physical," between the "material" and the so-called "non-material," "social," etc., which is today perhaps the chief obstacle to the unity of all the sciences, including the social. Futile controversies of all kinds which now engage our time could be relegated to limbo to keep company with their scholastic predecessors, if we but scrutinized the language in which these profound propositions are couched. What should be required in this field?

In the first place, I would require a demonstrated knowledge of the nature of symbols and signs and the nature of

[3] For elaboration of an important aspect of this matter, see L. Bryson, "Writers: The Enemies of Social Science," *Saturday Review of Literature*, October 13, 1945.

the rules governing the use of symbols in human communication, including the ability to write coherently and unambiguously, especially in the forms of exposition and argument. This I take to be what Hutchins meant by his emphasis on grammar, rhetoric, logic, and mathematics. I thoroughly agree with him, *provided* it is clearly understood that I mean not dead languages, primitive logic, or obsolete mathematics, but *modern linguistics, semantics, and the mathematics and logic of the past fifty years.* The *evolution* of all of these may be studied properly and profitably as history, mythology, magic, numerology, and superstition. But first, and as *necessary equipment* for the study of this background, the student must be familiar with contemporary knowledge on these subjects.

This familiarity with the basic means of communication must be insisted upon as a primary requirement of a general education because it is necessary not only to all intelligent day-to-day adjustments but also to the acquisition of scientific knowledge.

IV

Objections to Revision

We have now reviewed three great defects of contemporary education: First, its failure to agree upon science as its central, unifying discipline; second, its failure to impart that discipline; and third, its failure to teach adequately the arts of denotative communication. In discussing the remedies for these defects, I have, in addition to emphasizing the teaching of science itself, also pointed out that while both physical and social science are dependent upon mathematics, the social sciences are in a considerable degree also dependent upon certain types of history, philosophy, and certain parts of many subjects usually included in that vague category called the humanities.

Let us consider now some of the principal objections to the revisions of the curriculum which I have proposed. First among them is the very genuine feeling in many quarters that what is customarily called the humanities would be grossly neglected under my proposal. As I have already shown, the humanities are not necessarily destroyed, but their traditional subject matter is identified and sorted out.

The traditional content of the humanities is, then, destined to be scrutinized and evaluated from the standpoint of its contribution either to science or to art. A great deal of this content, as I have already indicated, becomes auxiliary to the various sciences just as, for example, mathematics and the history of physics are regarded as auxiliary to that science. The rest of the content of the humanities, or such part of it as can make the grade, will be assigned to the domain of art. When this process is completed, we shall find that the vital content of modern education can best be classified, as I have said, into two main categories, namely, science and art.

I am not under the delusion that this classification, like all others, is anything more than a construct of man's convenience. Human life and activity is a whole, and all our classifications of it are merely devices to facilitate discussion by taking up one aspect at a time. I am aware how inextricably art and science are interlaced at all stages, and how indistinguishable they are in their higher reaches and in their deeper meanings. But in their lower and middle registers they can be distinguished by sufficiently objective criteria to make the classification useful. Symbolic behavior and the use of symbols primarily to evoke in others the feeling and imagery experienced by oneself is the artist's main concern. The use of symbols in so standardized and formal a way as to denote, depict, or portray behavior in an impersonal and verifiable manner, and stripped as nearly as may be of the communicator's personal feelings about the matter communicated—this is the scientist's ideal. It is the difference between an artist's picture, whether represented in oils or in words, and an accurate blueprint or a photograph. When this and other differences that could be mentioned are recognized, there is no misunderstanding and no conflict. The reason that the battle today centers on the status of the humanities is precisely that they are a no-man's land, an undefined and uncharted terrain littered with the intellectual wreckage of the ages as well as with some choicest fragments of that wreckage. Neither science nor art is interested in destroying that content which has value, but in classifying and salvaging it.

On the other hand, this reclassification on a functional basis of a conglomeration of historic residues, would expose and demand justification for a multitude of curious fragments and lore which today are accorded undue prestige because they have been hitherto the identifying mark of the gentleman, and of the privileged class. Now, there is nothing that man clings to with such tenacity as the marks of his social status. We may, therefore, expect the most determined resistance in certain quarters to the type of reclassification I propose. Researchers in the field of radio programs tell me that one of the reasons for the great popularity of the "Information Please" programs is that every time a humble listener knows an answer he temporarily basks in the company of the intellectually elite, who, the inference is, know all these answers. Crossword puzzles serve in part the same purpose. As amusement, I have no objection to these intellectual scavenger hunts, but I object to confusing this sort of thing with liberal education.

The periodic exposés of the appalling ignorance of American history is pretty much in the same class. To the extent that the test shows how the schools fail to teach what they are trying to teach, the results are, of course, of interest. To the extent that the results are an index of our ignorance of other things that really matter, the test may be of some importance. But I am unable to become greatly excited over the results until someone shows me exactly why knowledge of most of the items called for is important. There is one thing that is worse than ignorance, and that is to know a lot of things that aren't so; or, what amounts to the same thing, to have exaggerated notions of the importance of what we know just because we know it, or because such knowledge has been the perquisite of the socially privileged in the past.

It would be well if each generation scrutinized pretty closely its intellectual heritage from this point of view, which is what I am here recommending regarding all subjects. I have not demanded that any of it necessarily be discarded, because, as is well known, both antique ideas and antique furniture can have great charm and yield much esthetic satisfaction, and a satisfying sense of the roots which we have in the past. All I have advocated, in effect, is that the antique chair be properly so labeled, lest someone mistake it for something else and sit on it, possibly to the ruination of both it and himself.

A second objection to the type of reorganization of the curriculum which I have proposed is that it would make education excessively preoccupied with the contemporary world, that we would be "forgetting human culture has traditional roots," etc. This seems to me highly absurd. Contemporary science is simply the tested residue of the groping of all the ages. Its roots lie in remotest antiquity. But why lead the student into all the fantastic blind alleys of man's history before he has achieved standards by which he can at least recognize a blind alley when he sees it? We recognize the absurdity of this in some fields. For example, we do not teach the chemistry, physics, and biology of ancient Greece or the Middle Ages *before* students are acquainted with present knowledge of these subjects; any one who seriously proceeds to talk and reason about chemistry, physics, and biology in the manner of the Middle Ages is regarded with amusement. But, in other fields, the same obsolete type of discussion is seriously considered, and is even supposed to be somehow more respectable than contemporary positive knowledge. A widespread social psycho-pathology is the result. As Eric Temple Bell has said: "Only an exceptionally skeptical mind can maintain its judicial disbelief in what is demonstrably false when a subtly fallacious argument is developed with all the superb skill of the old masters, who did not know they were lying, but who nevertheless lied with incomparable genius."

This is also all I have said about history. We cannot escape history, and we do not want to escape it, because all events that have occurred at all are *ipso facto* history, including yesterday's chemistry experiment. You will notice, however, that the other sciences, in reconstructing the history of the physical universe, have illumined their subject more through the knowledge of contemporary science than through the perusal of ancient documents. The history of the physical universe consists largely of what physicists today say *must* have happened in view of what we know *can* happen and *is* happening. Human history on the other hand, we seem to feel, is more like some kinds of cheese and wine—the older it is the better it is. This will change as the social sciences advance.

Preoccupation with ancient books seems to have had a strange fascination for certain types of minds. One problem in this connection has never been successfuly clarified. On the one hand it is contended that the way to profound and original thinking is through the study of "the" classics. Yet the authors of the original classics somehow obviously made the grade without the benefit of *"the"* classics. The inference seems clear that, however valuable "the" classics may be, they are not essential to the development of the type of minds held up for reverence and as models by the classicists, since "the classics" themselves were not available to their authors.

How can this be? Jefferson's account of his own case is suggestive. Regarding the origin of the Declaration of Independence, he said, shortly before his death in 1826:

"We had no occasion to search into musty records, to hunt up royal parchments, or to investigate the laws and institutions of a semi-barbarous ancestry. We appealed to those of nature and found them engraved on our hearts."

But perhaps a mistake has been made in including the thoughts of Jefferson among those of classical caliber. Consider, for example, Jefferson's estimate of a leading classic in his own field of government. In a letter dated July 5, 1814, to John Adams, Jefferson wrote:

"...I amused myself with reading seriously Plato's Republic. I am wrong, however, in calling it amusement, for it was the heaviest task-work I ever went through. I had occasionally before taken up some of his other works, but scarcely ever had patience to go through a whole dialogue. While wading through the whimsies, the puerilities, and unintelligible jargon of this work, I laid it down often to ask myself how it could have been, that the world should have so long consented to give reputation to such nonsense as this? ...But fashion and authority apart, and bringing Plato to the test of reason, take from him his sophisms, futilities and incomprehensibilities, and what remains? ...His foggy mind is forever presenting the semblances of objects which, half seen through a mist, can be defined neither in form nor dimensions. Yet this, which should have consigned him to early oblivion, really procured him immortality of fame and reverence...The doctrines which flowed from the lips of Jesus himself are within the comprehension of a child; but thousands of volumes have not yet explained the Platonisms engrafted on them; and for this obvious reason, that nonsense can never be explained."

In short, no one has any quarrel with classics as here understood. Indeed, the whole argument of this book is for more and better classics regarding human behavior. It is our contention, however, that only as a result of long and laborious social research can these classics be written. In the meantime, to be sure, we have to get along as best we can with what we have. But all seem to agree that, when it comes to managing social relations, we are not doing conspicuously well, although "the" classics venerated at present have been at our disposal for centuries. At the same time it is recognized that our really outstanding achievements, namely, those in the physical sciences, are based not on ancient "classics" but on the output of contemporary laboratories.

V

Educational Organization and Techniques

Having now dealt at considerable length with the content of a general education, and the objections to the changes I have suggested, it would be logical to turn to the subject of teaching and educational administration. To enter upon this vast and nebulous field, however, would be going beyond the scope of the present essay. We must confine ourselves to a few general observations.

We shall doubtless continue to make increasingly adequate provision for education in the conventional ways we have followed in the past. That is, we shall build better schoolhouses, train teachers better, pay them better, etc. But if we are to make headway in what Wells called the race between education and catastrophe, we shall also have to develop new and more effective techniques of reaching the masses of people than we have employed thus far. Fortunately, technological developments already point to ways in which this can be done. Radio, television, movies, and phonographic reproduction are surely destined for extensive educational uses under the supervision and as part of our general educational system.

The lower schools will doubtless continue to put much emphasis on personal and classroom instruction, because the process of socialization of the individual in the early stages, at least, must always involve direct association. As we approach the college and university, however, the reliance must be increasingly on the written rather than the spoken word. In fact, if we could teach people, first, how to read, and second, develop in them a continuing curiosity about the world, the educational problem would be largely solved. Libraries and directors of reading and laboratories would still have to be provided. But, with these available, the knowledge of this world is at the disposal of him who can read, and who has the incentive to do so. As for that large group who cannot learn to read by any means now known, we should abandon the attempt to teach them subjects which require reading. We depend entirely too much upon methods of teaching which have been out of date since the wide diffusion of printed matter became possible. It is a waste of time to listen to a lecture for an hour, the content of which a properly trained person should be able to get much better in fifteen minutes through reading. Yet the habit of relying on lectures persists, first, because it is a habit, and second, because most people have not yet learned to read with ease and accuracy. The fact is, of course, that most people at present respond more readily to ideas communicated orally by a person on a stage because it is fundamentally a dramatic setup with all the appeals of that form of communication. Still, as a method of simple exposition, it is a primitive method, wasteful of time, and it must be increasingly supplanted by reading.

Why is it so hard to teach people to read? Eight years of elementary school is largely devoted to it; four years of high school devote much attention to it; and yet people come to college and graduate therefrom still lacking the power and/or the inclination to absorb knowledge from the printed page. Correlative, of course, is the need to teach those who have material to communicate to write it so that it can be absorbed by a minimum of effort on the part of the reader. But whoever succeeds in arousing an enduring curiosity in men *and* discovers how to teach them to read will have done more to revolutionize our education than all others put together. This is the all-pervading, the staggering need in the pedagogical aspect of educational reconstruction.

If people could really read, a system of correspondence study, especially in adult education, could reach the entire population. The radio may also achieve this result but with a relatively large waste of time, for the very reason that much important material cannot be satisfactorily communicated orally. That is one reason why we have blackboards, maps, and charts in our lecture rooms. Television may solve this problem. There remains, however, the fundamental handicap that oral communication is relatively slow.

A word should be said in this connection about round tables, panels, and discussion groups which are so popular and from which so much is hoped, especially in adult education. Their strength lies chiefly in their dramatic and entertainment aspects. This is important, in view of the appalling neglect of education for leisure and recreation generally. But the fact that entertainment may be an important means and accompaniment to learning should not cause us to confuse it with learning, for amusement caters chiefly to the emotions and prejudices we already have, whereas learning frequently asks us to abandon our most cherished beliefs. Too frequently the popular "town hall" program is very little more than an occasion for people to assemble and engage in moblike demonstrations in support of the prejudices they already hold. I am not opposed to these meetings any more than I am opposed to other sports, and I do not deny their value especially as entertainment and as training in the etiquette of public meetings. I merely warn against expecting too much from them as serious education.

In general conclusion, then, the principal problems confronting education are these: How can we, as soon as possible, inculcate into our population (1)a rudimentary understanding of what is the nature of scientific method as applied to human affairs, and (2) a conviction that this is the only effective approach? Ultimately we must reach the grade schools and the high schools, because only through them can the bulk of the population be reached. But we can reach the lower schools only through teachers, and only teachers who are themselves familiar with the scientific viewpoint can be expected in the long run to shape both the curriculum and the teaching of the lower schools in that direction. I would begin, therefore, with a more adequate training of teachers in scientific method regardless of what subject they are to teach. A more thorough and a different kind of social science than is now taught is a second requirement. Together with these changes must go our adult education program calculated to achieve similar changes in the general public so that they will support and demand the relevant changes in the schools.

Finally, as the fountainhead of all education there stand the laboratories of research. If teaching is not to degenerate into mere pedantry and if our schools are not to become mere retailers of dead and possibly erroneous knowledge, research in every field must go forward with that eternal exploration and testing through which alone the truth is found. Only to the extent that the headwaters are constantly refreshed and replenished can the streams that trickle into every hamlet and home yield sustenance and solace and hope.

CHAPTER V

The Arts, Literature, and the Spiritual Life in a Scientific World

I

Science as a Threat

A GREAT many people are sympathetic to the views expressed in the preceding chapters. They are impressed by the achievements of science in the fields where it has been tried. They are intrigued with its possibilities in the management of human affairs. But they have the uncomfortable feeling that the advent of science in the latter field will somehow result in destroying some experiences and moments of life to which life seems to owe its highest values. These experiences are frequently associated with esthetic, artistic, and spiritual matters. They consist of these aspects of life that have justly been celebrated in song and story, in poetry and in music, in painting, in the dance, and in other arts, and in religious experiences. Clearly, if the advancement of science in the human sphere is to be at the expense of these most cherished aspects of man's experience, then this is too high a price to pay for the things that science can give us. The question deserves, therefore, the most careful inquiry.

As an illustration of extreme thinking on the possible effects of science upon man, we may cite the views of the late Victorian novelist George Gissing:

"I hate and fear 'science' because of my conviction that for a long time to come if not forever, it will be the remorseless enemy of mankind. I see it destroying all simplicity and gentleness of life, all beauty of the world; I see it restoring barbarism under the mask of civilization; I see it darkening men's minds and hardening their hearts; I see it bringing a time of vast conflicts which will pale into insignificance 'the thousand wars of old', and, as likely as not, will whelm all the laborious advances of mankind in blood-drenched chaos."[1]

As another example of this view, we have the German philosopher Schelling protesting against "that blind and thoughtless mode of investigating nature which has become generally established since the corruption of philosophy by Bacon and of Physics by Boyle and Newton."[2]

There are perhaps few writers, artists, or clergymen who would take such extreme views today. It is true that the controversy over evolution which flared up, I suppose for the last time, in the early 'twenties, frequently went to considerable lengths in its denunciation of science. But the storm soon died down. It was already too late to campaign successfully against science. The history of man's resistance to new ways of thinking and to new ways of doing things is sufficiently familiar. Most of the important advances of natural science and technology have encountered this opposition. Yet it is now quite generally agreed that these developments have been to man's advantage, especially as applied to the physical world.

These introductory remarks should not be taken to imply that contemporary novelists or philosophers, or even the clergy, on the whole, take a hostile attitude toward science. On the contrary, many of the warmest supporters of science are to be found in the ranks of literary men, artists, and philosophers. It should not be forgotten either that poets

[1] Morris Goran, "The Literati Revolt against Science," *Philosophy of Science*, vol. 7, no. 3 (July, 1940), p. 379. I have drawn on other material also from this article in the following paragraphs. See also Lyman Bryson "Writers: The Enemies of Social Science," *Saturday Review of Literature*, October 13, 1945.

[2] *Ibid.*

have found in science and the scientific attitude liberation for the spirit as well as the body of man. The contemporary literary attitude, in so far as it is critical, is perhaps more truly represented by Joseph Wood Krutch when he writes:

"We are disillusioned with the laboratory, not because we have lost faith in the truth of its findings, but because we have lost faith in the power of those findings to help us *as generally* as we had once hoped they might."

This is apparently what also bothers Robert Hutchins when he dramatically points out, in a statement to which we referred in the first chapter, that "the world has reached at one and the same moment the zenith of its information, technology, and power over nature and the nadir of its moral and political life." Perhaps many people have hoped that the chemistry laboratory would provide a plastic, a drug, or an elixir which would obliterate racial and cultural differences, and which would transform intolerance, envy, and greed into tolerance and brotherly love. In this hope they are doubtless doomed to disappointment. The results they seek must not be expected from the laboratories of physical science. The knowledge of how to improve human relations can come only from social science.

The preceding chapters have attempted to show that this is a reasonable hope, and many people already agree that this is our most promising prospect. But they are not clear, and they are properly concerned, about the possible effects on the arts and religion if we adopt the scientific approach to human affairs. Above all, they want to know about the relation of science to values and the ends of all our striving. The present chapter will deal with these questions.

II

Science and the Esthetic Life

In fairness, the opponents of the point of view I have taken in this book should be classified in several categories, because they do not present a united front or agree as to the grounds for opposing the inroads of science.

There are, first of all, those who have vested interests in the traditional sources of authority which science threatens to supplant. This is the familiar story of the conflict between the wizard and the scientist, the medicine man and the physician. The politician and the journalist are likely to ridicule and denounce the social scientist as a "brain truster" when the latter gets in their way. Philosophers, historians, and literary and classical scholars also sometimes regard social scientists as intruders. The old-fashioned priest cannot help noting that the psychiatrist, the psychologist, the sociologist, and the social worker are intruding on the traditional preserve of the clergy as healers of the mind, the soul, and the spirit. The enlightened clergyman, of course, welcomes these technicians and adds them to his staff. In addition, he himself takes training in these fields.

The expansion of knowledge and the resulting specialization further complicates the situation. Time was when the great philosopher was also the great poet, the great astronomer, and perhaps also the great physician and the great soldier. Today the development of each field does not permit many of us to excel in more than one or a few subjects, and few people aspire to the varied attainments of a Leonardo or even a Sir Philip Sidney. But the feeling survives in some quarters that the poet, the philosopher, the novelist, and the classical scholar, rather than the social scientist, are still the authorities on human relations.

As we have seen, science does not depreciate or contemplate the abandonment of the contribution of any of these fields. But during the transition, social science seems to encroach upon traditional, vested areas in the academic world. To the extent that this is the basis of the negative reaction, the problem is the same as in the case of vested interests in any other field.

A second class of critics of science are offended at the ugly, unethical, and unhealthy accompaniments of technology and industrialism. These results may include such matters as losing a job or a business through technological developments, occupational disease, regimentation, the destructiveness of modern war, etc. This group of critics feel that science and its creation, the machine, have devaluated, standardized, and mechanized human life. They accordingly advocate return to the farm, the village, and the handicraft shop. To these critics the scientists can only retort that the conditions complained of are rather the result of failure to apply science also to human relations, and that science cannot be justly indicted for the various misuses to which it is put and for some of the by-products which it creates. As for the proposal to return to a pre-industrial age, all that needs to be said is that actually no considerable number of people will support the suggestion because it amounts to a proposal to starve to death a substantial part of the present population and greatly to reduce the standard of living for the rest.

Finally, a third group, closely related to the one just discussed, consists perhaps mainly of the artist, the literary person, and the theologian who resent what they feel is the effect of science in reducing and derogating man's stature as the lord of creation and as a spiritual being. The attitude is largely an emotional one, but represents a very genuine feeling that the position I have taken contemplates denuding life of its highest and finest moments. Certainly, unless we can meet this objection, our case for science in human affairs is lost.

The critics here under consideration resent the impersonal, detached, and dispassionate methods of science. They crave a more personal communion with nature through feeling and sentiment. Their attitude is expressed in such lines as the following by Walt Whitman:

> When I heard the learn'd astronomer:
> When the proofs, the figures, were ranged in columns
> before me;
> When I was shown the charts and the diagrams,
> to add, divide and measure them;
> When I, sitting, heard the astronomer, where he
> lectured with much applause in the lecture room,
> How soon, unaccountably, I became tired and sick:
> Till rising and gliding out, I wandered off by myself
> In the mystical moist night air, and from time to time,
> Look'd up in perfect silence at the stars.

The implication which some people seem to get from these lines and hundreds of others that might be quoted is that the scientist, poor earthbound creature that he is, can never take a walk in the "mystical moist night air" and look up in perfect silence at the stars. Or at least, if he looks, he will not vibrate properly. What is more, this group of critics apparently is worried lest the progress of science will also prevent everyone else from looking with proper emotional appreciation at the stars.

It is a little difficult to deal with the objections of this group because it is hard to know exactly what it is that worries them. But much inquiry among those who are worried leads me to the following conclusions:

First, it appears that man always has been deeply stimulated emotionally by the mysteries of life. From this fact some of our friends seem to be worried lest, as a result of scientific investigation, the supply of mystery should give out and leave us omniscient but terribly unhappy and without any esthetic enjoyment. If so, perhaps they will be reassured by Veblen's statement to the effect that the net result of scientific investigation is to make two questions grow where one grew before. Incidentally, and in so far as fundamental mystery is also a requisite for religious emotion, this should also be reassuring to those who wonder whether science is not out to destroy religion.

A second type of worry seems to be merely that as scientific sophistication increases, man's appreciation of esthetic, ethical, literary, and artistic things necessarily decreases. I know a botanist who takes his students into the woods and fields when the foliage is at its most gorgeous peak and discourses enthusiastically about what he sees. His students greatly enjoy these walks and admit that their attention has been called to many things they previously did not see. But I have also heard them express the view that their instructor, alas, must be a very unhappy man, because, if one knew as much as he does about the names and life habits and insides of all the plants, one's esthetic enjoyment of them would presumably be largely destroyed.

In the same institution is a physicist who gives special courses in color and in sound, because of his lifelong interest in painting and music. At one time he delivered extemporaneously a considerable discourse on the physics of an especially gorgeous sunset that happened along. There was widespread sympathy expressed in the college at the poor man's evident inability to enjoy the sunset instead of analyzing it. For myself I can testify that I have on numerous occasions been asked by troubled students whether the matter-of-fact analysis of human relations and emotions does not greatly decrease one's enjoyment of the finer things of life. The significant thing is that the question is always about somebody else's specialty, rather than your own.

Ask the musician if his study of harmony and counterpoint has decreased his enjoyment of music and he is likely to go to the other extreme and assure you that without these studies you can't *really* enjoy things musical. But he may feel quite concerned about the esthetic life of our physicist and almost certainly about the esthetic experience of the botanist and the sociologist.

It is true that scientists, in common with other people, including laborers, merchants, bankers, and poets, sometimes become so preoccupied with what they are doing that they do not develop exactly well-balanced personalities. That is hardly a peculiarity of the scientist. It is probably as frequently true of artists and religious devotees. Also it is true that the rigors of scientific education frequently do not leave as much time for other things as one might wish. On the other hand, the artist is inclined to overlook the profound esthetic and artistic aspects of such pursuits as science and mathematics, as well as the high literary and artistic appreciation of many prominent scientists.

Altogether, may we not dismiss as naive and groundless the fears about the deleterious effect of science upon esthetic appreciation? In fact, I think we may even go a bit farther and suggest that this type of objection amounts to little more than an unconscious or at least unspoken provincial ethnocentrism or conceit which enables people to cherish the conviction that they themselves, their education, and their outlook represents the ideal type and model for humanity. There are people who are quite sure that he who has not read this particular poem or that particular classic has not partaken of the higher experiences of life and must necessarily be a stunted and underprivileged person as compared with those who have imbibed of a common curriculum, preferably at some exclusive school. Thus "humanists" compile lists of characters from novels and solemnly assure us that not to have known these particular figments of somebody's imagination is "to be imperfectly alive." The fact, I suspect, is that the highest emotional, esthetic, and ethical experiences may be achieved in a great variety of manners, and, as at least one poet has pointed out, even God fulfills Himself in many ways. Perhaps it would be well to keep that in mind when the old order changeth and science continues to advance.

So much, then, for certain general aspects of the question we have been considering. Let us now turn to some more specific issues.

In the first place, the fear that the extension of science, especially its development in the social sphere, will somehow be at the expense of the arts seems especially unwarranted on the basis of past experience. What have been the facts in that respect thus far? As science has advanced, the arts have profited, at least quantitatively, in every way. The masses of men today have access to the best artistic products in music, painting, and literature on a scale incomparably greater than at any other time in history. A far greater percentage of the population today earns a livelihood in these pursuits and a far greater percentage has leisure and education to enjoy artistic products. I know there is argument among artists as to whether the quality of the product is equal to that of other times. That is a question which can be decided only in the perspective of history and, in any case, must be left to those capable of judging the quality of artistic products. In the end they will probably be judged by the permanency and generality of their appeal. But the advancement of the social sciences as contemplated in this book could, in any case, only improve the situation for the production and consumption of all the arts, for, favorable as the present situation is, when compared to previous ages, including the grossly exaggerated "golden ages," the arts are still step-children in our culture.

The advancement of the social sciences will not only make this clear, but will also provide the means for the elevation of the arts to a more prominent place. Science is not an end in itself, but the most effective means so far discovered by man, for whatever ends he chooses to pursue. This complex of ends, or goals, has always been designated as the "good life," however the content and meaning of that phrase may have changed throughout the ages. The significant thing to note is the permanently prominent position of the arts in an otherwise changing conception of the good life. If, therefore, man adopts a more effective means for achieving all his ends, the arts should be one of the first and main beneficiaries of such a development. In short, the arts can perhaps come into their own only in a scientifically ordered society, which will provide the technical means, the health, and leisure with which to pursue the arts both on the creative and the appreciative levels, on a scale hitherto impossible.

But why, then, do we have the present agitation on the part of literary men against the encroachments of science? I have already reviewed some of the more general reasons. Literature, as contrasted with the other arts, seems to feel itself especially threatened, because it is regarded by many practitioners as something in addition to being an art. Throughout history, and to a considerable extent at present, literature has been regarded also as a sort of social science. In fact, from where do the overwhelming majority of people today draw the principles of sociology, psychology, political science, and economics which become the guide to their conduct and to their estimate of human affairs? Some of the main sources are unquestionably the novel and its dramatic counterpart, the movies, plus the newspaper columnists. To the extent that these representations coincide with or symbolize the scientific truth, this is of course admirable, and represents indeed one of the great and proper contributions of literature and the other arts. Unfortunately, it is not yet clear to many writers that, in the event that the portrayals of literature do not check with scientific fact, science must take precedence as a guide to all practical adjustments. If that conclusion is depressing to the literary fraternity, they need only to realize that one of their most important functions is to expound, advocate, and diffuse scientific thinking and findings.

This is generally recognized in all questions involving the physical sciences. The trouble seems to be that many so-called writers have not yet heard of social science of the type I am talking about. They confuse psychology with psychoanalysis, economics with capitalism, communism, or some other ideology, and sociology with everything from Christian socialism and social work to transcendental philosophy. A social scientist can still get the best-established scientific facts refuted at any literary cocktail party by citations to the contrary of characters and situations in current novels. If one objects to the possibly imaginary or at least unrepresentative character of the cases cited, the objection is promptly met by citations from additional novels, with perhaps a movie and a quotation from the leading newspaper columnist thrown in to clinch the point.

One can hardly find fault with artists for this confusion as to whether science or art is the arbiter of objective fact in the modern world as long as prominent professors of social science are themselves not clear on the point. Thus, a teacher of political science at Cambridge University, writing in the *New York Times Magazine* on "Europe's Portrait of Uncle Sam," makes the following revealing statement:

"The movies do give an impression of ease, of a social fluidity that may be false but is certainly seductive. I have read too many powerful American novels to assent readily to the thesis that, on the whole, life in America is easier for the common man than in any of the European industrial countries, but were it not for those powerful novels I should assent to this view, and most Europeans only know and believe what they see in the movies."

That is, while most people in Europe get their ideas about America from the movies, this "social scientist" is not so gullible. He checks on the reliability of the movies by reading novels. In short, if we really want to know how things are at the bottom of a rabbit hole, read a work by Lewis Carroll for the report of an eyewitness named Alice. Or, if you want to check the reliability with which the English novel portrays English life, visit the English movies.

Undoubtedly we have here the source of much of the current confusion and controversy over the place of science, literature, and the humanities in our educational system and as a guide to social action. It is regrettably true that social science has not yet made itself sufficiently convincing to many literary men as well as to some philosophers, preachers, and the public generally to cause them to take the findings of social scientists as more reliable statements of fact than are the impressions of artists. In this attitude, the artist and the public is, of course, partly justified by the inadequate, not to say ludicrous, character of many studies reported by social scientists. This is the more true because the defective sociological study is more likely to secure wide diffusion than the more competent work. Thus, when humanists investigate sociological research in order to appraise its validity relative to Thomas Aquinas, they choose studies that lend themselves to adverse criticism. This is a favorite method among the literati, and necessarily so, since many of them cannot even read the better social science monographs. Under the circumstances, it is impossible to blame them for their conclusions.

I cheerfully agree that, in the absence of social science, literary and artistic products of all kinds doubtless do afford some clues and approximations as to what may be facts of general validity. I agree that proverbs represent perhaps the earliest attempts at generalizations of the type at which science also aims. It is also probably true that the Bible, as scientists have pointed out, represents, broadly, the science of the periods when it was written. Indeed, it is one of the artist's and humanist's greatest trumps to be able to point to some poem or novel or philosophic work which, hundreds of years ago, enunciated some principle which has been finally corroborated by laborious scientific investigation only in our day. What is overlooked is that in these same literary works are a hundred other generalizations totally without scientific foundation of any sort. There is no doubt that throughout the whole range of literature there are numerous cases in which the artist's intuition has subsequently been validated by science. Unfortunately, what we must know, and what we don't know before scientific testing, is *which* of the artist's numerous intuitions are reliable. The prestige of the Greek philosophers rests in a large measure upon precisely this phenomenon, namely, the capacity of people to remember the philosophers' good guesses and to forget the fabulous "boners" of which they were also guilty.

It should be clear, then, that literature and the other arts are not substitutes for social science. However, they can be most valuable to science in two extremely important ways: In the first place, the arts, and literature especially, can serve as a source of scientific hypotheses. In the second place, the arts are invaluable in communicating and dramatizing scientific truth and in general and through emotional appeal can motivate to action in accordance with what is scientifically sound and possible. The arts can be, and have been, guilty of corresponding mischief when their presentations do not correspond with scientific truth. They can then set nation against nation and involve vast masses in futile and disastrous quests which can result only in frustration and disappointment. Not the least of the evil by-products of war is the prostitution of the arts in the interests of falsehood.

In addition, of course, the arts have important functions of their own in the emotional and imaginative development of man, and in the enrichment of experience with which I shall deal a little later. I am speaking here only of the contribution of the arts to science as the accredited arbiter of physical and social fact.

It appears, then, that the assumed conflict between science and the arts is entirely without valid foundation and is based on a misconception of the functions of each. I am not under the delusion that in this brief space I have fully described these functions. They have been quite fully and, for the most part, quite adequately described by workers in different fields.[3] I have here merely tried to indicate the nature of their interrelationship and the general basis upon which their respective functions may be determined.

III

Science and Religion

When we turn to the subject of religion and its probable status in a scientific world, we encounter at the outset the difficulty inherent in any attempt to discuss something which for most people has not been adequately defined. Yet the nature of the discussion will depend entirely upon this definition. If, for example, we accept Einstein's view that science itself can be the religion of the devoted scientist, then obviously no further discussion is necessary of

[3] See T. C. Pollock, *The Nature of Literature: Its Relation to Science, Language and Human Experience* (Princeton, N.J.: Princeton University Press, 1942). John Dewey, *Art as Experience* (New York: Minton Balch and Co., 1934). C. P. Snow, *The Two Cultures and the Scientific Revolution* (Cambridge, England: Cambridge University Press, 1959).

the place of religion in a scientific world.[4] Also, we find distinguished and accredited clergymen taking essentially the same view in declaring that any devoted pursuit of an aspiration is religion. Thus, Harry Emerson Fosdick has declared from his pulpit that even nazism was a religion, albeit a bad one. Professor H. R. Mussey has shown rather convincingly, I think, that Russian communism, especially in its first decade, fulfilled all the requirements of a religion as generally conceived. What is more, if one analyzes the essential components of religious behavior as set forth in the ablest anthropological literature I have been able to find on the subject, namely, that by Durkheim and by the late Edward Sapir, one is impressed by the soundness of this broad definition of religion. If so, there is clearly no problem of science *versus* religion, especially since some of these religions propose to employ and rely completely upon scientific methods in the attainment of their goals.

There are those, however, who will object to definitions so broad as those suggested above. They will point out that while those definitions may be defensible, they merely avoid the real issue in which people are interested, namely, what will be the status of their particular religion in a scientific world. This is obviously a very different question, involving as it does all the factors pertaining to the survival of a particular institution in a particular culture, as contrasted with a general behavior trait characteristic of all cultures. In short, what most people really hope will be discussed, when they turn to the subject of religion and science, is approximately this: If the social sciences advance as we have advocated in the preceding chapters, what

[4] See A. Einstein, "Science and Religion," *Science News Letter,* September 21, 1940, p. 181. Also E. F. Haskell, "The Religious Force of Unified Science," *Scientific Monthly,* June, 1942, pp. 545–51. Also S. Brody, "Science and Social Wisdom," *Scientific Monthly,* September, 1944, pp. 203–14.

might be the effect on that system of beliefs and practices constituting, for example, a particular subdivision of the Christian religion as of 1960 in the United States? In view of the fact that less than half of the human race even profess the Christian faith, and that millions of the rest hold religious beliefs which involve no god or other supernatural beings, this is clearly shifting from the general subject of religion and science to a pretty specific aspect of it. Nevertheless, it is in this specific form that it touches individuals, and the issue should therefore not be avoided by indulgence in mere generalities.

To start with, the question that is uppermost in the minds of many people in our culture is this: What about God and what about free will in a scientific world? Fortunately, we can answer this question satisfactorily, for all practical scientific purposes, without becoming involved in the metaphysical issues which traditionally becloud the subject. That is, for the practical scientific purposes that concern such a scientific world as I have outlined, you may hold whatever views you like regarding both God and free will. All I need to point out and establish for scientific purposes is the great regularity and predictability with which men *will* things. I can predict the will and the choices of men by exactly the same techniques I use to predict other natural phenomena. The same may be said about God. He is clearly a being with remarkably and demonstrably regular habits. It is this regularity and predictability of the will of both men and gods that is of interest to science.

I do not say that various metaphysical questions upon which scientists and theologians differ may not have significance from other points of view. Nor am I predicting what the changes in these beliefs may be in the future either under scientific or non-scientific conditions. I merely point out that the study and prediction of regularities in social behavior and the formulation of laws from such data can go forward without settling any metaphysical questions of ultimate causation with which science is definitely not concerned. For the same reason, scientists should not become involved in questions having to do with other worlds, if any, than that which is within reach of their methods. Social scientists should perhaps take more pains to make this clear, in order to achieve more enthusiastic cooperation on the part of that public which may prefer one rather than another of the various possible metaphysical positions.

There is no reason why all should not agree, as indeed they do in the physical sciences, regarding the demonstrable validity of the methods of science in arriving at laws describing regularities and probabilities of any behavior whatsoever. Why not (as I shall argue in the next chapter) leave the metaphysical questions to the taste, the temperament, and the needs of the individual? Scientists have themselves argued regarding the relative merits of the corpuscular theory and the wave theory of light, to say nothing of views as to what makes the electrons dance. These are important questions from some points of view. But these differences in no way affect either the established laws governing these phenomena or practical engineering enterprises.

The importance of this view regarding the widespread controversy about science and religion cannot be too strongly emphasized. We are in danger of cutting ourselves off from the most powerful tool that man has yet invented for the achievement of his wants (*whatever* they may be), because of the absurd belief that we cannot take advantage of what science has to offer in social as well as in the physical predicaments without first settling certain metaphysical questions which actually are not relevant to the problems that confront us. This is the more absurd because we already have worked out a *modus vivendi* on the subject in the physical sciences. Essentially the same physics, chemistry, and biology is taught in institutions maintained by the church as in those maintained by the state. These sciences are the same in Christian as in non-Christian lands. If the advancement and application of physical science should have had to wait upon the solution to everybody's satisfaction of all the metaphysical questions and cultural preferences of all peoples, then penicillin, electricity, aviation, and the radio could never have gained general usage. The capacity of physical science to command the respect and allegiance of the most disparate races and cultures and creeds is precisely its unique and priceless attribute. This characteristic, this general validity and applicability of science entitles social as well as physical science to the support of all men, however they may differ on other matters. That consideration should be pondered not only by the clergy, but also by those scientists who mistakenly try to make out that only under their own culture, their own petty political, economic, and religious creeds, and their own form of government can science prosper and be of service.

So far as the ethical, esthetic, and recreational appeal of religions are concerned, there is likewise no reason why they cannot continue with even greater freedom in a world in which social as well as physical science is dominant. Ethical norms will change in the future as they have in the past, and human institutions, including the church, will accommodate themselves to change in the future as in the past. A developed social science will greatly facilitate this adjustment, because through science man can secure a very much more adequate knowledge of the consequences of different types of conduct, instead of relying upon ancient and arbitrary authority for this counsel. The esthetic and recreational aspects of religion might be expected greatly to increase in a scientifically ordered society.

I use the word recreation in its literal sense to designate those behaviors in which man has always engaged because of the satisfactions which they yield in themselves rather than as means to other ends. They are the activities to which man has always looked for rest, recuperation, refreshment, rejuvenation, rebirth. Religious exercises, rituals, and devout observances generally have always appealed, I suspect, far more deeply to this basic craving than have the changing cosmologies, the metaphysics, the ethics, and the exhortations which are a part of some religions. This recreational craving is as ancient and deep-seated in man as his other basic hungers, and for this reason alone religious behavior might readily hold its own or even increase in a scientifically ordered society. Religion, in this respect, is in the same position as literature and the other arts, which can reach their full and proper place only in such a society. The tragic neglect of adequate recreation in contemporary society, and the probable resulting increase in nervous and mental disorders and criminal conduct is one of the most scandalous aspects of a society with resources and technology which, if scientifically ordered, could provide leisure and opportunity for the spiritual development of man through all the arts, including religious ceremonials, to a degree never before approached in history.

By the spiritual development of man, I mean nothing more mysterious than the development of his powers of communion, communication, imagination, sympathy, and empathy through his capacity for symbolic behavior by all and every means whatsoever. Man's spiritual life consists of his thoughts, usually in the form of words, and other symbolic imagery. We mean by such words as "purposes" and "ends" the word pictures that man conjures up from his desires and his longings. These word pictures or other imagery, when recorded, are as objective phenomena as the pictures that hang on our walls. We can study the processes by which man creates them, and we can study their effects upon people, including their creator, just as we study the effects of other environment.

This understanding of the "mental," "spiritual" life in terms of man's own symbolic behavior (his thoughts, dreams, imagination, aspirations, worship, etc.) in no way depreciates it or denies to it any of its exalted, precious, or sublime nature. After all, it is a remarkable thing that the eons of evolution should finally have brought forth creatures that can exchange thoughts and feelings with one another and with imagined beings through speech and artistic expression. The flights of which the human mind (spirit, soul) is capable are a staggering and awe-inspiring phenomenon. It is not surprising that man sought the explanation of this power outside of nature. The spirit of man is not less remarkable or less sacred when it is seen as the crown and the spire of nature.

The symbolic behavior of man has been referred to throughout history by such words as mind, soul, and spirit. The phenomena to which these words refer have been variously explained through the ages. As science advances, it can abandon the words and change the explanations. But it cannot and has no desire to ignore the phenomena. It is in this sense and for this reason that religion and spiritual phenomena are as proper subjects for scientific investigation as any other aspects of man's behavior.

IV

Science and Ethics

We have attempted to show that science, as the most effective instrument of achieving *whatever* goals we may pursue, should be welcomed by all men, however they may disagree on questions of final causation, the origin or the final destiny of man. In short, we merely obey here the ancient injunction to render unto Caesar the things that belong to Caesar. The more general form of that adage is this: Render unto science the things that belong to science and to metaphysics the things that belong to metaphysics.

There is nothing, then, in the proposals of this book or in the social order which it envisions which cannot be supported by the adherents of every religion, every political and every economic faction interested in the efficient and economical achievement of whatever objectives they pursue. This is, of course, precisely what exasperates all provincial and ethnocentric groups. They want science for the propagation of their own narrow cults alone. The critics of the position I have here advanced are especially enraged over the fact that I propose to ask the masses of men themselves what they want and reserve for science merely the modest function of telling men how to get their wants, and the costs and the consequences thereof. The critics, including those who worry most about whether or not my position is democratic, are sure that it is the privilege of science, by which they mean themselves, to tell the rest of mankind what they should want. As a result they become involved in hopeless contradictions, rationalizations, and hypocrisies over the question which appears so crucial and inscrutable: Can science tell us what objectives to pursue, what we should want?

We have already touched on some aspects of that question in Chapter II and we are now in a position to complete that discussion. In the first place, the ends that individuals have pursued throughout history are a matter of historical fact. Pronouncements about the ends man *should* pursue are usually called matters of opinion and matters of taste. But opinions and tastes, as such, are also matters of fact. In the framework of science, taste and the choice by which we evidence it is the result of all the experience of our species through all the centuries of its evolution, plus all the conditionings of our own generation and our present surroundings. In all the talk about objectives and values, it is interesting to note that the most important fact, namely, the remarkable *agreement* in the human race upon the *principal* ends of human striving, is rarely emphasized. Our common humanity indeed determines the more general of these objectives.

What are some of these principal ends upon which there has been almost unanimous agreement? There has always been remarkable agreement that physical survival, security, and a livelihood for the individual and for the group are desirable ends. It has also been agreed with remarkable unanimity throughout the ages that satisfying group association, activity, and growth for its own sake of the type I have classified broadly as recreational, in which I have included artistic, religious, and spiritual experiences, are also highly valued ends in themselves. Note that I am talking here not about what ends I think man should pursue, but about the ends that he has in fact pursued through the centuries. Collectively, he has pursued these ends because he is the kind of organism that craves these satisfactions and finds them worthwhile. For any scientist, preacher, or poet to say under these circumstances that man *should* pursue other ends, simply amounts to saying that man *should* be a different being than he is, that he *should* have had a different evolutionary history than he has had. This may be a proper indulgence for preachers and poets but hardly for scientists.

It may be pointed out that in every group there are individuals or minorities, or possibly even majorities, who disagree with the goals which their culture as a whole pursues. They may even disagree regarding the very generally accepted goals I have just mentioned. That is, they may declare this life is a vale of tears, that recreation is a sin and a delusion, that nirvana or heaven is the proper goal. This is clearly a question of temperament and taste or conditioning. More frequently it is merely declared that many of the specific activities engaged in will compromise or defeat the more ultimate or desirable goals. This is a question of fact. Questions of taste may be allowed to individuals and groups as nearly as is compatible with community standards. A so-called free society usually also allows people of different tastes the privilege of trying to convert one another, while insisting in the meantime on conformity in those respects which are set by whatever mores are accepted or enforced. Questions of fact, however, immediately come within the dominion of science. And, as we shall see, most *questions even of taste and ethics in the end turn out to be determined by what people believe to be facts.*

It is in this connection that we discover the true relation of science to values, to ethics, to ends. As we have seen, there is general agreement by the masses of men on the large and broad goals of life as evidenced by man's behavior. Everywhere he tries to keep alive as best he knows how, he tries to enjoy association with his fellow creatures, and he tries to achieve communion with them and with his universe, including his own imaginative creations. The sharp differences of opinion arise about the *means,* the *costs,* and the *consequences* of different possible courses of action. Here science is the accredited method. As we have seen in Chapter II, what science has done in the fields in which it has been applied and can do regarding human society is precisely this: It can acquaint people reliably with (1) the *possible* alternate courses of action, and (2) the costs and consequences of each course. *Whatever* people do under these circumstances will constitute their "valuing"—their Values. Values represent, therefore, no special or unique problem in the social sciences, as is often asserted. All behavior, human and sub-human, may be regarded as evaluative, and it is as regular and predictable as the behavior of the inanimate world.

Even when the alternate possible courses have been laid before a community and the costs and consequences of each have been pointed out, there will still be differences of opinion because of different tastes and temperaments as to which course we should pursue. This must be decided, as it has always been decided, by whatever method of consensus happens to be accepted and accredited, or at least happens to be operative, in the group. These methods vary all the way from purest democracy, through the various types of aristocracy and authority, to so-called absolute dictatorships. I say *so-called* absolute dictatorships because in the end even the dictator does not fly in the face of the most generally accepted underlying mores of his people. For myself, I happen to prefer the democratic method of determining consensus of opinion and sentiment. I accordingly rejoice, as I noted in a previous chapter, in the development of instruments for opinion and attitude measurement which, if properly applied, could perhaps restore the vitality of the relatively pure democracy which has fallen into considerable disrepute because of the obsolete technology through which it seeks to function. If future generations develop still other mechanisms of achieving communality, I certainly will not insist that they blindly adhere to the techniques to which I am habituated.

The reader may be disappointed and even offended at the suggestion that a question so enshrined in oratory and emotion as is the problem we are discussing should be answered in such mundane fashion as is here proposed. With all due respect to the ponderous intellects and the even more ponderous tomes that have wrestled with the subject, I make bold to suggest that the net wisdom so far bequeathed to us is contained in a passage not usually invoked in the present connection:

"Cheshire-Puss," she [Alice] began . . . "Would you tell me, please, which way I ought to go from here?" "That depends a good deal on where you want to get to," said the Cat.

Can *science* tell man what *direction* he should go? Yes, if man will tell scientists where he wants to go. *Should* man then do what he collectively *wants* to do? Assuredly, unless you wish to set up a higher source of wisdom and authority than man's collective experience through the ages, interpreted and supplemented by the reliable prediction of scientists as to the costs and consequences of different possible courses of action. To declare under these circumstances that man *ought to want* something else than what he does want is merely a semantic trick by which the speakers seek to invoke "ethical principles" derived from some source outside of nature, in support of what he himself wants man to want. This is also the answer to those who claim that the universities are shirking their responsibility if they do not tell man what he should want. Those who presume to set up such higher authority above and beyond the most competent appraisal of man's experience should equip themselves with more convincing credentials than they now possess, if they wish to avoid the suspicion that they are merely setting up themselves as the authority which they claim to derive from authorities outside of nature or of experience.

As a child of nature, man has from the beginning striven, however inadequately, to pass on his experience to succeeding generations. Those experiences which he has considered to his advantage he has cherished as moral and good conduct—the things he *should* do. Those experiences which he has considered to his disadvantage he has held up as immoral and bad—things he *should not* do. In the mores of every culture these guides to what man should or should not do are enshrined. Sometimes they are celebrated and inculcated through elaborate institutional and ritual practices and are attributed to supernatural sources. But the source of all has always been sovereign man himself and the nature of which he is a part. What he has lacked hitherto is a reliable method of observing, recording, generalizing, and interpreting human experience. As a result, he has frequently saddled himself for centuries with absurd and fantastic notions of his own history and experience, with resulting distortion of "values," mores, and institutions. Science provides a reliable method for a correction of these distortions.

It is true, then, that science does not directly tell us what we should want. But it does not follow that science cannot deal practically with human values. By providing reliable estimates of the near and remote consequences of alternative courses of action, science *conditions* the choices—the values—of men. Only by subjecting the artist's and the seer's visions to this evaluation can we avoid the frustration and neuroses which are today the result of vain hopes, unfounded beliefs, fantastic aspirations, and the pursuit of mutually exclusive goals.

It will always be the privilege of the poet to smash the world to bits and then rebuild it nearer to the heart's desire. Even poets have recognized, however, that a desirable condition for this enterprise is first to possess some grasp of "the sorry scheme of things entire." With or without such grasp, it will still be the privilege of philosophers, poets, and seers to conjure up their own visions of the good life and to invite others to join them in quest of it. It is gratuitous to assume, however, that there is any such thing as *the* good life for all times, places, and conditions of men. Perhaps only the totalitarian-minded have the conceit necessary to so grandiose a presumption. These aspirations do not spring full-blown from the mind of seer or saint, nor are they deduced by sheer logic from revealed premises. Visions of the good life spring from man's experience through the ages. Social science is the most effective instrument for determining reliably what that experience has been and what it means in terms of attainable present goals. Science not only frees, stimulates, and disciplines the imagination of men, but it provides the means of telling us what we most want to know regarding our visions: Which of our aspirations are achievable and at what cost? Which ones are mutually incompatible? It is the lack of a reliable method of answering such questions that results in controversy, frustration, and despair.

CHAPTER VI

How to Live with People Who Are Wrong

I

The Passing of Religious Wars

WE HAVE reviewed in the foregoing chapters some of the proposed approaches to the problems of associated living. Our conclusion is that the general method of attack called natural science, applied to the *social* as well as to our physical predicaments, is the most promising. But even if the scientific developments outlined should come to pass, it must be admitted that there would still be a great variety of differences among individuals and groups on philosophies of life—religions, values, tastes, and "world outlook."[1] Accordingly, we have to recognize the very probable continued existence of a highly pluralistic world. Some of these variations will be tolerated and even encouraged as providing variety, diversion, and esthetic satisfaction in an otherwise highly standardized culture. But other deviations will appear to be monstrous and intolerable, or, at least, simply *wrong*. Can science contribute anything to the technique of living with people who are "wrong"?

The nature of the problem is reflected in a recent statement on the floor of the United States Senate. In a resounding speech on foreign policy one senator said: "What we need to say, as American People, and say in unmistakable terms, is that we are right."[2] Doubtless millions of Ameri-

[1] Dr. Rogers Williams, in *The Human Frontier* (New York: Harcourt, Brace and Co., 1946), reports at length on the tremendous variability of tastes, traits, and other characteristics of human beings.

[2] Senator Thomas J. Dodd, *Congressional Record*, February 26, 1959, p. 2725.

cans find it reassuring to know that in an uncertain world and in our quarrels with the perverse representatives of other countries, whatever happens, we are right. The problem, therefore, as we define it at present, seems to be (*a*) how to induce our opponents to agree that we are right, and conversely, that they are wrong; or (*b*) how to live in the meantime as amicably as possible with people, who, to us, are wrong. As regards the first alternative, and in spite of occasional "thaws" in the cold war, there appears to be no reasonable prospect in the foreseeable future of a change of viewpoint on either side. Accordingly, whether we like it or not, it appears that both parties are destined to live for some time with people whom they consider wrong. This chapter explores some possible improvements in the techniques of such living.

The situation itself is not unfamiliar. Nearly all of us have had experience in living with people whom we consider wrong. Most family quarrels and community disagreements arise and have to be endured because people consider each other wrong. What makes the present impasse in foreign relations so critical is our reluctance to utilize even the traditional techniques that have been invoked successfully in such situations during the whole of recorded history. Such techniques are usually called compromise, forbearance, or toleration. At their best, these processes result from a more careful and accurate appraisal of the situation as a whole. Such reconsideration seems to have been involved in the abandonment of religious warfare.

After many centuries of religious wars, the *de facto* leaders apparently came to the conclusion, reflected in the Peace of Westphalia (the treaty ending the Thirty Years War), that it is possible to live quite peaceably with people who are considered completely wrong in their most fundamental beliefs. We need not become involved in the question of whether the professed religious grounds were in fact the "real" reasons for the Thirty Years War. What is probably indisputable is that seventeenth-century Europeans were weighed down with the most ironclad prohibitions of any form of compromise with sin, or with people who held false beliefs. Nevertheless, the secular princes managed to get together on a sort of permanent *modus vivendi* regarding religious issues which avoided the moral incongruity widely felt today at any suggestion of compromise on political and economic ideologies and ways of life.

We have not been able to achieve, as yet, in the field of politico-economic ideology and practice an adjustment of the type which fortunately has been achieved within the domain of religion. By making compromise equivalent to "appeasement" and identifying the latter as a flagrant immorality, or at least a tactic of proven futility, we deprive ourselves of the historic techniques of dealing with situations exemplified by the cold war. In short, history seems destined to report that the leading statesmen of the twentieth century were less skilled in the technique of living with politico-economic deviants than with religious dissenters.

This is not to imply that the shift from conversion by the sword and the general approach of the Inquisition to the religious *modus vivendi* was not bitterly contested in the time of transition. The then incumbent Pope, with plenty of Protestant support, denounced the Peace of Westphalia which he declared "null and void, a curse, and without any influence or result for the past, the present, or the future." There was much clamor in religious circles for the resumption of the Holy War. Nevertheless, both Catholic and Protestant princes were sufficiently impressed with the economic burden and general futility of ideological war (even without airplanes and atom bombs) to abstain from further holy wars and, for the most part, from ideological wars in general.

Of course, wars continued over boundary disputes, territory, trade, colonies, and other tangible considerations which adjustment to a changing world doubtless requires from time to time until more effective and economical means of settlement may be developed and adopted. Also, many countries and communities have, down to the present time, suffered bitter internal struggles over religious doctrines. But the international wars of the past two centuries were not primarily of an ideological character. It remained for the United States of America in 1917, under the moral lash of Woodrow Wilson, to reverse our historic policy as laid down in the Monroe Doctrine and to return to the highly moral plane of religious war. The Democracy *versus* Autocracy-Communism controversy is clearly of this character. Some examples of the state of mind underlying the dominant contemporary statecraft, and its similarity to the theoretical orientation underlying the medieval religious wars will clarify the above allegations.

II

Theistic National Policy

Writing in *Harper's* for December, 1956, on "Woodrow Wilson among His Friends," Raymond B. Fosdick speaks with admiration of his former teacher as follows:

"Wilson was a deeply religious man. He believed that God was working out his purposes in this world, and once he made up his mind that a particular course of action represented the will of God, nothing could shake him loose from it. 'God save us from compromise' he used to say; 'Let's stop being merely practical and find out what's right.' When I saw him, a few weeks before he died, we discussed the League of Nations. With tears rolling down his face he said, 'You can't fight God!' To him the underlying principle of the League of Nations represented the fulfillment of a preordained purpose, and if he took any pride in the situation at all, it was that he had been an instrument, however faulty, in carrying out the will of God."

It may be objected that Wilson was a devoted Presbyterian, that the quoted statement was made when he was old and ill, and that under the growing secularization of the church such beliefs have declined during the last thirty years. Let us then look at some examples from the current scene. Consider first the following extract from an address by a recent cabinet member, a man in his prime and in full possession of his considerable faculties, a former law professor, affiliated with the Methodist Church:

"If man is to judge himself competently, the standards he applies to his conduct must of necessity be beyond his power to modify or define. If man has the power to define what his standards are, then they almost inevitably become what he wills them to be. Thus through man's infinite capacity for rationalization, many lies may seem to be truth, deceit may wear the cloak of honor, oppression may be practiced in the name of justice. This, to me, is the great ethical error of materialism, humanism and all other systems of philosophy which do not recognize the independent existence of absolute moral and spiritual values. . . . These values are spiritual and absolute, rather than material and relative . . . quite above and beyond the sphere of human development. It is for man to perceive these values as the lasting, immutable works of his God. *He must not conceive them as the property of his own mind,* to be twisted and distorted to suit the demands of expediency. . . . Thus we assert our conviction of *the divine source of a set of absolute spiritual values,* and with it we express a conviction that these values must not be subverted by any man. . . ."[3] [Italics mine.]

Now, in what respects do the lines quoted above reflect a state of mind comparable to that which underlay the religious wars of three or four centuries ago? It would be im-

[3] R. B. Anderson, *Time,* October 26, 1953, p. 27.

possible to improve on the answer given by Sir Oliver Franks, the former British Ambassador to Washington (1948–52). In his eulogy of John Foster Dulles, Sir Oliver said:

"Three or four centuries ago, when Reformation and Counter-Reformation divided Europe into armed camps, in an age of wars of religion, it was not so rare to encounter men of the type of Dulles. Like them, in vigorous and systematic reflection, he had come to unshakable convictions of a religious and theological order. Like them, he saw the world as an area in which the forces of good and evil were continuously at war. Like them, he believed that this was the contest which supremely mattered.

"This is not just a fanciful analogy. I am sure that John Foster Dulles believed that he had been called to be Secretary of State at a time when the world was again divided into armed camps by moral beliefs and metaphysical doctrines. It was in this light that he conceived the struggle between Communism and the Free World. He saw international political issues in moral terms because in the end he saw them as theological. . . . For such a person the business of thought and action is not a tentative exploration by trial and error of what is expedient; it is a deductive exercise which, by applying known principles to the facts, shows how to move to the pre-established goal. In my conversations with him, I thought Dulles's mind essentially worked in this way. It was for these reasons that he did not depend very much on the advice of the State Department." (*London Times,* May 31, 1959.)

Perhaps the significant sentence in the above quotation so far as the present essay is concerned is the remark that "For such a person the business of thought and action is *not* a tentative exploration by trial and error of what is expedient." [Italics supplied.] Nor even, one may add, a scientific estimate of the costs and consequence of different possible courses of action. For such a person, in other words, there is possible only a Wilsonian declaration *against* "compromise" and *for* "what is right." And how does one know what is right? Secretary Anderson has answered unequivocally and with admirable honesty: "It is for man to *perceive* what is right as the lasting and immutable works of his God."

This is still undoubtedly the keynote of present foreign policy of the Western world, especially of the United States. However, the results of the policy since we seriously began to attempt to implement it (in 1917) have been conspicuously disappointing. What alternative can be offered?

III

Social World Picture

To merit consideration, an alternative policy should first of all provide a method and a technique of securing agreement on important questions without resorting to war.[4] In the second place, an acceptable alternative policy should not demand too great interference on either side with the individual's spiritual, ideological, aspirational predilection—what is usually called a person's "philosophy of life." What methods and instruments have been developed over the centuries that appear to fulfill the above requirements to a high degree in the fields where these methods seriously have been tried?

One of the most significant things about the highly developed physical sciences, such as chemistry and physics, is their remarkable capacity to compel agreement on answers to physical and chemical questions. At the same time, scientists allow themselves the widest imaginable liberty and difference of opinion regarding ultimate values and philoso-

[4] See Anatol Rapoport, *Fights, Games, and Debates* (Ann Arbor, Mich.: The University of Michigan Press, 1960).

phies of life. Consider in this connection the sharply differing views of three scientists of great distinction.

First and mostly briefly, representing one extreme, we have Bertrand Russell declaring with his customary clarity:

"I am myself a dissenter from all known religions, and I hope that every kind of religious belief will die out." (*The Will to Doubt,* p. 17.)

Second, in a somewhat intermediate position, we have the words of Albert Einstein:

"The more a man is imbued with the orderly regularity of all events, the firmer becomes his conviction that there is no room left by the side of this ordered regularity for causes of a different nature. For him, neither the rule of human will nor the rule of Divine Will exists as an independent cause of natural events. . . . In their struggle for the ethical good, teachers of religion must have the stature to give up the doctrine of a personal God. . . . Whoever has undergone the intense experience of successful advances in this domain of scientific thought is moved by profound reverence for the rationality manifested in existence. By way of understanding, he achieves a far-reaching emancipation from the shackles of personal hope and desires, and thereby attains that humble attitude of mind toward the grandeur of reason incarnate in existence, which, in its profoundest depths is inaccessible to man. This attitude, however, appears to me to be religion in the highest sense of the word." [5]

Consider finally the following position of Felix Mainx, author of the monograph on biology in the *International Encyclopedia of Unified Science,* of which Bertrand Russell is a member of the Advisory Committee:

"By 'World Picture' (Weltbild) and 'Philosophy of life' (Weltanschauung) we shall here understand well defined domains of human spiritual life. By 'World Picture' will be

[5] Albert Einstein, *Ideas and Opinions* (New York: Crown Publishers, Inc., 1954), pp. 48–49.

meant the total representation of the world which we can form on the basis of all statements of empirical science. . . . On the other hand by 'Philosophy of life' is to be understood the totality of all the emotionally tinged strivings and ideas of a man, the totality of his value judgments, of his ethical and esthetic maxims. . . . It is the conviction of the author of this monograph that *every* philosophy of life rests on a faith, on a decision to trust, which can only be reached from an inner human experience. . . . The synthesis between philosophy of life and world picture, as a spiritual task of every time and of every man is therefore never only a matter of empirical science, but always a matter or faith. The author of this monograph, who counts himself fortunate in having as a Catholic Christian, a positive belief, has always felt this consistent synthesis to be the greatest fulfillment of the spiritual life." (Felix Mainx, "Foundations of Biology," pp. 82–83.)

Is it possible that the gulf which separates Eisenhower and Khrushchëv (or for that matter, Karl Marx and Thomas Aquinas) is any greater than that dividing Russell and Mainx? Yet both of them, as well as the less extreme Einstein, subscribe to a common method in giving us a reliable picture of our physical world. As a result of such agreement regarding the world picture, it is possible for all three to reach a considerable measure of agreement regarding the probable costs and consequences of what man can and may do in such a world. These costs and consequences may then be calmly weighed in the balance of people's respective value systems. Whatever choice we make under these circumstances will at least be a deliberate and calculated action.

Such agreement on a method of reliable estimation of probable consequences of man's action is exactly what is lacking regarding political, economic, and social questions. For example, Germany appears not to have expected a two-front war. American hopes to make the world safe for democracy were not realistic estimates of the consequences of our participation in World War I. Our miscalculations of the consequences of World War II need no recapitulation. One of the reasons for these errors was that we did not have (and do not have) a common *social* world picture nor a fully or generally accepted method of arriving at one.

I have not said that agreement on a social world picture necessarily would result in agreement on social *goals* any more than a common physical world picture has resulted in agreement on physical goals. Agreement on the method of producing the atomic bomb did not at all produce a unanimous judgment on the part of the scientists concerned upon whether to drop it on Hiroshima and Nagasaki. The scientists did agree that the physical results of dropping the bombs would be essentially as predicted and subsequently observed. All I have said is that a common and accepted scientific method permits the development of a verifiable theory of human society (social world picture) and makes possible a realistic estimate of the results of different possible courses of action. Foreseeing these costs and consequences might in turn inhibit the adoption of a given policy, the results of which can be shown in advance to be disappointing from the point of view of the value system we cherish.

IV

The Behavioral Sciences

For an indefinite period, it seems certain that we shall have to live with people whom we consider wrong in their thoughtways, and in some of their other major behavior patterns. We are likely to continue to deplore certain types of political and industrial organization and value hierarchies different from our own. What is needed is some arrangement which will permit relatively peaceful coexistence in spite of our differences. We have advanced the idea that the instrument of such a *modus vivendi* regarding the physical world has been found in physical science. The rapidly developing social and behavioral sciences must ultimately play a similar role in human social relations.

This conclusion is predicated on a development of the behavioral sciences comparable to that of the physical sciences. Only the future can finally demonstrate the validity of this predication. Suffice it to say that the behavioral sciences are making considerable headway and are undoubtedly here to stay—and to develop. Some of the developments which may have revolutionary impact are already on the horizon. Consider, for example, the recent invention of the fabulous computers which make it possible for one man to solve in a few minutes problems which without the machines would require a hundred years. Granted that technological improvements cannot of themselves advance science, few will question the importance to the physical and biological sciences of improved microscopes, telescopes, and all the paraphernalia of the scientific laboratory. Less well known, but of tremendous significance, is the recent development of improved techniques of precise measurement of attitudes, values, and public opinion. Or consider, for example, the possible significance of "game theory" in the social sciences. The basic conditions in any game (in game theory) include such factors as (1) risk and uncertainty, (2) conflict of interest, (3) statements of *all possible alternatives,* and (4) weightings of these possibilities according to *the preferences of the players.* These factors may also be the crucial ones in many non-game social situations. Impressive new developments in "decision theory," "information theory," and "communication theory" are also under way.[6]

Many of these more promising leads are highly abstruse

[6] See R. D. Luce and H. Raiffa, *Games and Decisions: Introduction and Critical Survey* (New York: John Wiley and Sons, 1957).

and have been hitherto explored mainly by advanced mathematicians. To suggest that such theories may produce answers to the question posed by the title of this chapter may seem like giving him who asks for bread not only a stone, but a whole truckload of gravel. The ways of science are notoriously involved and laborious. Much basic scientific work looks to quite a distant future for its practical results. No wonder that the simple declaration that "we are right" seems a less bothersome answer.

Among the common objections to any solution which looks beyond the immediate present is that, as one much-admired economist put it, "In the long run we are all dead." An even more famous expression of the same idea is, "After me the deluge." It is doubtless true that in the long run "we" are all dead, but it is precisely a function of the social sciences to point out and deal with the fact that while individual statesmen and economists are (perhaps fortunately) all dead in the long run, human society has continued to exist for hundreds of thousands of years and is likely to continue —atom bombs notwithstanding. National and international policies designed only for the lifetime or term of office of the incumbent officials cannot be expected to be very effective even in the short run, as the cold war so clearly shows. Day-to-day adjustments will be necessary under any policy. But successful national policy cannot consist *only* of "play-it-by-ear," daily stopgap measures. It is not a matter of "waiting" for the social and behavioral sciences to develop. It is a matter of developing the behavioral sciences as rapidly as possible *while* we make the best day-to-day adjustment we can.

V

Technique of Cooperation

We have distinguished sharply two types of questions, (*a*) those that deal with preferences, aspirations, and hopes, and (*b*) those which deal with verifiable, predictable facts and sequences of events. Doctrinaire communism or capitalism are examples of the former type. That is, they are philosophies of life or preferences not arrived at by scientific procedures. A mature science of economics dealing with the probable cause-and-effect relationships in production, distribution, and consumption of economic goods would be an example of the second type of question. We have attempted to show that wide disagreement on philosophies of life is compatible with quite peaceful coexistence and cooperation provided there is agreement on the methods and conclusions regarding the facts of empirical science. The so-called ideologies which so agitate us currently may turn out to have very little to do with the practical business of living in a community.

This is not to question the importance of philosophies of life and ideologies to the adherents thereof. As is well known, ideologies of the most fantastic sort have for considerable periods of time sustained people under the most trying circumstances. We have merely pointed out that wide disagreements on philosophies of life are compatible with survival and with peaceful community living, whereas certain other questions regarding, for example, public safety, health, population increase, and subsistence in a highly specialized industrial civilization require a minimum of recognition and conformity by everyone.

Is this separation of two types of questions historically defensible? We have already referred to the consequences of the abandonment of large-scale religious wars in the Western world. The Peace of Westphalia made no compromise regarding ideology, nor did it prohibit the respective parties from trying to proselyte and convert the world to their own way of thinking. The treaty merely stipulated certain practical arrangements in international relations, while leaving ideological issues severely alone. Thus it became permissible for members of different religious factions to observe the more usual amenities of community living, which in time developed into the present practice of rather cordial overt relations between members of different religious denominations.

This moderate cordiality does not imply any compromise of doctrine. When the Catholic Church periodically invites other Christian sects to return to the one and only true faith, they do so in a friendly and compassionate spirit, but they make it unmistakably clear that, after all, the Protestants are wrong. The Protestants usually receive this invitation and its implication with politeness and tolerance but point out that unless the Catholics abandon certain errors of doctrine, unification cannot be considered. In the meantime, both continue to work together for community objectives on which they agree, and overt hostility is highly exceptional. Each faction proselytes in a mild manner, and the missionary work of each goes forward with some success at the other's expense. Ideologically, both declare as a matter of course or as an implicit postulate that it is their ultimate purpose to conquer the world, not by force, but by gradually convincing everyone that does not subscribe to the true doctrine that they are simply wrong. Why should not situations like the cold war be similarly adjusted?

We have employed the contemporary world picture of physics and chemistry to implement our material preferences; the results far transcend those achievable under the alchemists' world picture. In the same way it seems legitimate to use the new social world picture which the behavioral sciences may provide in implementing our ideological and spiritual preferences. Now, if these preferences happen to be world peace, harmony, and justice among nations, *a more highly developed technique of cooperation and compromise* could make all the difference between achieving or not achieving our preferences.

In short, our present blundering may be regarded as a confession of technical incompetence in the management of human affairs rather than as attributable to malice and perversity on the part of our opponents. To remedy this technical deficiency is the purpose and the proper province of the social and behavioral sciences.

This view is apparently not entirely discounted even in circles where we might least expect it to appear (albeit inadvertently). Writing in *Foreign Affairs* (October, 1959), Nikita S. Khrushchëv said:

"Is it possible that when mankind has advanced to a plane where it has proved capable of the greatest discoveries and of making its first steps into outer space, it should not be able to use the colossal achievements of its genius for the establishment of a stable peace, for the good of man, rather than for the preparation of another war and for the destruction of all that has been created by its labor over many millenniums? Reason refuses to believe this."

The significant note in this statement is that the "colossal achievements" referred to are those of *generally accepted science* (which will some day include the behavioral sciences) —East and West—not the results of the turgid conjurings of Karl Marx (with or without the Lenin-Stalin dispensations). When Western leaders make a corresponding acknowledgment regarding their private revelations that "we are right," we shall be under way in the business of living with "people who are wrong."

CHAPTER VII

Conclusion

CAN SCIENCE save us? Yes, but we must not expect physical science to solve social problems. We cannot expect penicillin to solve the employer-employee struggle, nor can we expect better electric lamps to illumine darkened intellects and emotions. We cannot expect atomic fission to reveal the nature of the social atom and the manner of its control. If we want results in improved human relations we must direct our research to the solution of these problems.

To those who are still skeptical and unimpressed with the promise of social science, we may address this question: What alternatives do you propose that hold greater promise? If we do not place our faith in social science, to what shall we look for social solutions? We reviewed in the first chapter the principal sources to which man has looked in the past. In the preceding chapter and elsewhere throughout the book we have described the dominant current faith as a moralistic-legalistic thoughtway sharply at variance with our dispassionate, analytical attitude toward "natural" (i.e., non-social) threats, problems, and disasters. The moralistic-legalistic method proceeds by classifying people at the outset as right or wrong, good or bad, aggressive or nonaggressive, peace-loving or war-loving, and so forth. From this simple viewpoint it is easy to generate a widespread belief in the possibility and imminence of international amity from such a well-intentioned instrument as the Charter of the United Nations. Without in any way disparaging the efforts and achievements of this organization, we must recognize its limitations in the light of the goals that have been set for it.

Nothing has been said in this book against "world organization" as such, any more than we have opposed schools, hospitals, and other organizations as means intended to achieve desired ends. We have merely been compelled to call attention to the regrettable fact that good intentions are not a substitute for good techniques in achieving either physical or social goals. Many people resent having this called to their attention in regard to projects to which they are emotionally attached. Good intentions seem intrinsically virtuous, and we feel we deserve some credit at least for having "our heart in the right place." Furthermore, the suggestion that good intentions can become operative only through efficient techniques, and especially through science, seems to deprive the masses of men of the privilege of participation in the great enterprise of social salvation. It is not generally recognized that the support of the scientific approach can be a more idealistic and emotionally rewarding participation in community life than a primitive moralizing about the evil ways of those who disagree with us.

Social organization is merely a means to the end that men seek, and world organization is no exception. It deserves not blind enthusiasm but careful scrutiny of particular proposals. To begin with, the phrase "world government," without specification of the extent, scope, and nature of the proposal is vague and misleading. To most people "world government" probably means a state of affairs more or less resembling that of the principal national governments now in operation.[1] But just as national governments can be of great variety as regards their centralized control over the lives of its citizens, so world government can mean much or little.

[1] The favorite model is perhaps the U.S.A. For an excellent analysis of the romantic nature of the analogy, see N. A. Pelcovits, "World Government Now?" *Harper's Magazine,* November, 1946, pp. 369–403. See also Warren R. Austin, "A Warning on World Government," *Harper's Magazine,* May, 1949, pp. 93–97.

After all, there is nothing new about world organization. Long before questions of *government* in the formal sense arose in the relations between states, there had developed a very elaborate set of specific mechanisms for the handling of *specific* relations without becoming involved in the crucial question of *enforcement by a super state or supernational authority*. Thus, there were in 1950 more than one thousand private, international *nongovernmental* organizations dealing with business and finance, communications, transportation, travel, labor, agriculture, art and science, the press, education, religion, social welfare, sports, law and legal settlement of disputes, and pursuit of peace.[2] Increasing contacts with other cultures makes international organization in such matters desirable in peace as well as in war.

It is probably true that current interest in "world organization" is overwhelmingly concerned with its possible effectiveness in avoiding atomic war. If the United Nations should succeed in this one objective, perhaps no one would be very critical of its other extensive activities. Some of these clearly require international action, others do not. For example, certain arrangements having to do with specific relationships, such as mail service and public health, have been notably successful in achieving support among diverse cultures. They arouse very little opposition because they regulate important functions that people recognize to be beyond the power of national organizations. On the other hand, proposals such as the International Bill of Rights (modeled chiefly on the American pattern) probably have little chance of immediate and wide-scale acceptance or effective practice in the foreseeable future. Whether an African chief is to be allowed a hundred wives or only one, which is currently the officially approved quota in the United States, may be a matter of interest chiefly to the particular tribe within which

[2] L. C. White, *International Governmental Organization* (New Brunswick, N.J.: Rutgers University Press, 1951), p. 325.

plural marriage is practiced. This question is radically different, for example, from international control of communicable diseases. The latter can be shown (*a*) to be practically impossible to control except by international measures, and (*b*) to have very disastrous consequences upon quite remote and "independent" tribes. Perhaps relative world peace is an objective more nearly resembling the control of communicable disease rather than the marital mores of "underdeveloped" African tribes.

If we take the liberty of comparing, for the sake of illustration, war as a scourge comparable in its effects to the ravage of communicable disease, one is at once struck with the difference in attitudes, approach, *and the amount of scientific knowledge* involved in curtailing disease as compared with diagnosing, preventing, or limiting war. "World organization" at present is somewhat like an elaborate hospital full of activity and well-meaning personnel, but without any scientific knowledge of *what to do* about particular disorders in human social relations. We noted in the first chapter that one of the principal obstacles to a scientific approach to social problems is that we think we already know the answers. Currently, one of the most popular of these answers is something called "world organization." To many people these words are merely new symbols satisfying the requirements of an old thought pattern of the moralistic-legalistic type. It is the new medicine, the panacea, the "one world" of brotherly love that men have imagined to be just around the corner, with all the troublesome questions of social technology left out. For centuries men have dreamed of lands running with milk and honey, and have looked to governments, religions, and other "organizations" for achieving this goal. It is seldom realized that mainly through the wearisome development of scientific agriculture has any progress been made toward the realization of the dream.

One of the most popular ideas at present is that if some

problem which we do not yet know how to solve in a local village or factory is projected on a world stage it would more or less solve itself. Nothing so disgusts the enthusiastic declaimer on international race and minority problems as the suggestion that he first find out the nature of such problems and demonstrate their solution in a single local community. Industrial problems, employer-employee conflict, unemployment? Why, what we need is fewer restrictions on international trade and an international labor organization. Indeed, even in a nation whose total foreign trade, in the most prosperous times, is less than 10 per cent of its total trade, leaders proclaim that we must go to war to protect that 10 per cent in order to avoid unemployment and catastrophic decline in its standard of living.

Now obviously, very few people are, in fact, opposed to world organization for the same reason that very few are opposed to peace, prosperity, and happiness. The only question is about types and methods of organization, which is precisely what needs careful scientific investigation. But such inquiry is regarded as sabotage by the proponents of particular idealistic schemes of world organization. We are warned against saboteurs, especially among "people holding high government office or chairs in universities . . . who indulge in the puerile excuse that 'the time is not yet ripe'."[3] The implication is that if we had only had a sufficiently vigorous campaign for aviation in the Middle Ages, and if as a result enough people had voted for it, Columbus could have done his stint in a stratoliner instead of in his pitiful ships. The scientist, unfortunately, would have to point out that the time literally was not ripe—the absence of certain developments in steel manufacture, internal combustion engines, etc., would have defeated the most "dynamic organization

[3] Emery Reves, *The Anatomy of Peace* (New York: Harper & Bros., 1945). Quotations in the following pages are from this book unless otherwise noted.

behind an idea." It is this sort of practical consideration that infuriates the "one worlders."

War itself is a symptom of social tensions of various sorts. The latter are found in innumerable smaller communities as well as among nations. We need to direct our attention to these tensions—the *causes* of war, crime, and strikes—rather than at the symptoms. "Outlawing" the symptoms will not cure the disease. The intensive scientific analysis which is required for the real solution of any important problem and which has, in fact, been the condition of whatever progress we have made, is impossible as long as people do not even recognize the nature of the problem. Faith in miracles of international organizations seems to have superseded to a large extent faith in the Second Coming. We seem to be hoping for the social counterpart of the Industrial Revolution, without any of the drab and undramatic and technological work upon which the latter depended.

There are no grounds whatever for expecting all wars of whatever sort suddenly to disappear from the earth under any organization. The most we can reasonably hope for is their gradual reduction. Increasingly large regional units *within which* relative peace exists are in fact emerging in spite of our determined efforts to prevent such integration in some places. The integration of the United States of America and its domination of the Western Hemisphere resulted in relative peace in this part of the world for a century and a half. The process is now under way in Russia and adjoining areas. Europe doubtless would have emerged as a unified relatively peaceful region decades ago except for outside interference with that process.

I am not here arguing for the necessity of perpetual war. I am calling attention to the possibility that *there are "natural" processes in the social as in the physical world* and that there may be limits to the liberties man can take with these processes if he wishes to achieve *successful* organization and peace.[4] Are there demographic, ecological, economic, and cultural principles to which social organization must, within limits, conform? Are there certain principles governing human relations that are fully as important as those known to govern the construction of bridges? If so, had we not better get busy formulating and applying these principles? Without such knowledge no world organization can be more than another blundering trial-and-error attempt like those that have gone before. In the absence of more efficient methods, trial-and-error attempts are also defensible. This book has not opposed them, nor has it prescribed another world scheme of its own. I have merely advocated that we apply the tested methods of science in approaching the problem.

"Idea" propagation without scientific knowledge of how to achieve concrete goals is pure sophistry. The world situation requires more fundamental treatment than another indignation meeting, some more Kellogg pacts, "summit" meetings, or a world "Charter," although all of these may have their place. Any world organization which neglects at the outset to make extensive provisions for *scientific research for its own guidance* is doomed to failure. What is to be gained by organization unless administrators *know what to do,* including how to secure international support for action which can be shown to be adapted to the ends sought? The most idealistic and well-intentioned administrators in the world cannot perform the tasks we expect of them without scientific knowledge of the kind today possessed by engineers and physicians.

Somewhat the same point may be made regarding the

[4] For elaboration of this viewpoint see my papers "Scientists in Wartime," *Scientific Monthly,* vol. 58, February, 1944; "Sociologists and the Peace," *American Sociological Review,* vol. 9, February, 1944; and "Regionalism, Science, and the Peace Settlement," *Social Forces,* vol. 21, December, 1942.

conclusion of some politicians and physical scientists when they seek to escape their problem by piously declaring that "the real problem is in the hearts of men." After centuries of repetition by the clergy, and others, the cliché has a comforting and convincing ring especially to the confused and the weary. This book has repeatedly pointed out that the "hearts of men" were in quite satisfactory condition for centuries as regards their desire for more facile communication and transportation and relief from war, pestilence, and famine. Relief came not with a change of heart but through a particular method of using their brains, namely, the methods of science. If the mere wishes of the masses of men were the determining factor, wars would have disappeared long ago. The truth is that what is in the hearts of men regarding war has not yet found adequate expression in scientific knowledge and techniques.

The chances of another industrial revolution resulting from the solution of the problem of atomic fission has given rise to a wealth of speculation as to its possible effects on the social order. Some seem to believe that this development will automatically create a "world order." The reasoning seems to be that since we shall badly *need* world organization, therefore it will suddenly come into being. There is no foundation whatever for the doctrine that need alone results in either physical or social invention. If need was the sole determining factor the airplane and sulfa drugs should have developed centuries ago. These inventions came only when research had developed the necessary material, technological, and scientific foundations. Some people have been justifiably skeptical about the success of world organization until the necessary "moral" foundation for it has developed. They now apparently believe that the development of an especially destructive bomb somehow provides the needed moral foundation. Aside from specific questions of present world organization, most students agree that atomic power might readily call for organization with a high degree of centralized control not only on a national but on an international basis. In view of the effects upon individual lives of the totalitarian regimes we have seen so far, this prospect cannot but cause widespread concern at least in a nation cherishing other ideals. In this dilemma, the development of the social sciences alone can save.

To put it bluntly, in the present state of development of the social sciences, centralized administration of large national or international societies can be carried out perhaps only by precisely the methods that have thus far characterized such regimes, namely, ruthless suppression of all opposition and wholesale starvation or other deprivation because it is not known how such costs of a comprehensive centralized government can be avoided. A leader, however admirable in ability and intentions, attempting to administer centrally a large society today is somewhat in the position of a pilot trying to fly the modern stratoliner without an instrument board or charts. That is to say, it cannot be a very smooth flight. If he succeeds at all, it will be at the expense of much wreckage of men and materials. Successful piloting depends directly upon the adequacy and accuracy of the instruments in the machine, the charts by which a course can be pursued or modified, and the training of the pilot to read both aright. Only as a result of the development of the basic physical sciences can a large modern airplane either be built or flown. Only through a comparable development of the social sciences can a workable world order be either constructed or administered. The appalling thing is the flimsy and inadequate information on the basis of which even a conscientious executive of a large state is today obliged to act.

It comes down, then, to this: Shall we put our faith in science or in something else? We have already answered that question as regards our physical problems. Once we make up our minds to do likewise regarding our social predicaments, we shall at least have made a beginning in a promising direction. This is the question which ultimately must be answered by everyone, but first by scientists themselves, by legislatures, by the Foundations, and by individuals who endow and finance research and education. If it is answered in the affirmative, then social research institutions will make their appearance, which will rank with Massachusetts and California Institutes of Technology, Mellon Institute, the research laboratories of Bell Telephone, General Electric and General Motors, not to mention several thousand others. For some time the sponsors of these enterprises devoted to physical research have been wondering if the solution to social problems does not lie in the same direction. When they undertake to support social research as generously as they have supported physical research, they will obtain comparable results.

Finally, a word should be said to those who find the methods of science too slow. They want to know what we shall do while we wait for the social sciences to develop. Well, we shall doubtless continue to suffer. Executives will continue to decide on the basis of guess and intuition and to mistake their own voices for the voice of the people or of God. The nations will doubtless continue to rage and the people to imagine vain things. Life went on also in the days before anesthetics, vaccines, and sulfa drugs. These days also had their immediate and pressing problems. A few people, however, devoted themselves to research which could not possibly solve the current difficulties, but which have transformed our world. We do not abandon cancer research because the patients of today may not be saved by it. We shall probably become much sicker before we consent to take the only medicine which can help us.

Many of the fruits of science, however, can be used to advantage while still in the process of development. Science is at best a growth, not a sudden revelation. We also can use it imperfectly and in part while it is developing. When we give our undivided faith to science, we shall possess a faith more worthy of allegiance than many we vainly have followed in the past, and we also shall accelerate the translation of our faith into actuality.

ACKNOWLEDGMENTS

Most of the material of this book was first brought together in approximately its present form as Walker Ames lectures at the University of Washington in the spring of 1945. It is a pleasure to acknowledge my obligation to the University and its Department of Sociology for the invitation to deliver these lectures. Parts of the material have previously appeared in *Harper's Magazine,* June, 1943, and December, 1945, and in *The Scientific Monthly,* October, 1941, and February, 1944. The new Chapter VI in the Second Edition previously appeared in part in *The Humanist,* 1960, No. 2. Permission to reprint some of this material is hereby gratefully acknowledged.

G.A.L.

Index